THE volumes of the University of Michigan Studies are published by authority of the Executive Board of the Graduate School of the University of Michigan. A list of the volumes thus far published or arranged for is given at the end of this volume.

University of Michigan Studies
HUMANISTIC SERIES
VOLUME XI

CONTRIBUTIONS TO THE HISTORY OF SCIENCE

PART II. NICOLAUS STENO'S DISSERTATION CONCERNING A SOLID BODY ENCLOSED BY PROCESS OF NATURE WITHIN A SOLID

THE MACMILLAN COMPANY
NEW YORK · BOSTON · CHICAGO · DALLAS
ATLANTA · SAN FRANCISCO

MACMILLAN & CO., LIMITED
LONDON · BOMBAY · CALCUTTA
MELBOURNE

THE MACMILLAN CO. OF CANADA, LTD.
TORONTO

PLATE V.

PORTRAIT OF STENO IN THE PITTI PALACE.

THE PRODROMUS

OF

NICOLAUS STENO'S DISSERTATION

CONCERNING A SOLID BODY ENCLOSED BY
PROCESS OF NATURE WITHIN A SOLID

*AN ENGLISH VERSION WITH AN INTRODUCTION
AND EXPLANATORY NOTES*

BY

JOHN GARRETT WINTER

UNIVERSITY OF MICHIGAN

WITH A FOREWORD

BY

WILLIAM H. HOBBS

UNIVERSITY OF MICHIGAN

New York
THE MACMILLAN COMPANY
LONDON: MACMILLAN AND COMPANY, LIMITED
1916

All rights reserved

COPYRIGHT, 1916,
BY FRANCIS W. KELSEY, EDITOR.

Set up and electrotyped. Published September, 1916.

Norwood Press
J. S. Cushing Co. — Berwick & Smith Co.
Norwood, Mass., U.S.A.

Paperback ISBN: 978-0-472-75202-7

PREFACE

THE task of preparing this translation of Steno's *Prodromus* has been lightened by the generous help of several of my colleagues in the University of Michigan. To Professor W. H. Hobbs I am indebted for suggesting the work and for reading the entire manuscript, as well as for contributing a *Foreword*. Professor E. H. Kraus read in manuscript the section dealing with crystallography, and Professor E. C. Case gave helpful suggestions in questions of palæontology. A point in physics was clarified by Professor W. D. Henderson.

To Professor J. B. Woodworth, of Harvard University, my thanks are due for permission to reprint the section entitled *The Interpreter to the Reader* from his copy of the *H. O.* version, and for verifying certain references. Mr. Bernhard Berenson, of Florence, kindly furnished photographs of the portrait of Steno in the Pitti Palace, and of Duke Ferdinand II in the Uffizi. Dr. Fr. C. C. Hansen, of the University of Copenhagen, generously sent a photograph of the portrait of Steno as Vicar of Schwerin. But from the editorial side my greatest debt, and one I have especial pleasure in acknowledging, is to Dr. Vilhelm Maar, of the University of Copenhagen, whose scholarly edition of Steno's *Opera Philosophica* has been of invaluable service. In addition to furnishing photographs of the portraits of Steno, Dr. Maar has given me, by letter, not only much information, but also warm encouragement.

The recent publication of a facsimile edition of the *Prodromus*, by W. Junk (Berlin, 1904), and of the text, by V. Maar (Copenhagen, 1910), has obviated the necessity of presenting the Latin text in connection with this translation.

I am under much obligation to Mr. William H. Murphy, of Detroit, whose generosity has made possible this publication in the Humanistic Series.

JOHN G. WINTER.

ANN ARBOR, MICHIGAN,
March 15, 1916.

CONTENTS

	PAGE
FOREWORD	169
INTRODUCTION:	
I. Life of Steno	175
II. The Writings of Steno	188
III. Bibliography of the *Prodromus*	194
IV. Selected References	202
TRANSLATION OF THE *PRODROMUS* WITH EXPLANATORY NOTES . .	205
ATTESTATIONS	271
EXPLANATION OF THE FIGURES	272
INDEX	277

PLATES

V.	PORTRAIT OF STENO IN THE PITTI PALACE	*Frontispiece*
		FACING PAGE
VI.	PORTRAIT OF STENO AS VICAR OF SCHWERIN . . .	184
VII.	REPRODUCTION OF ORIGINAL TITLE PAGE	194
VIII.	REPRODUCTION OF FIRST PAGE	196
IX.	REPRODUCTION OF STENO'S FIGURES, 1–13 . . .	272
X.	REPRODUCTION OF STENO'S FIGURES, 14–19 . . .	274
XI.	REPRODUCTION OF STENO'S FIGURES, 20–25 . . .	276

	PAGE
REPRODUCTION OF TAILPIECE (p. 76 of Original Edition published at Florence in 1669)	270

FOREWORD

The Science of the Prodromus of Nicolaus Steno

In reading the *Prodromus* of Nicolaus Steno one should remember that the essay was written near the middle of the seventeenth century, when scientific observation was hardly thought of. All knowledge concerning the causes of natural phenomena was generally supposed to have been given by God directly to man, and the message was strictly guarded by the church. Giordano Bruno, who denied that there had been a universal deluge, and who had brought forward evidence that a change had taken place in the distribution of land and sea, was burned at the stake for heresy. More than a century later, and a half a century after Steno wrote, de Maillet, in order to express his conviction that the rocks of the earth were marine deposits, thought it necessary to disguise his name in the anagram *Telliamed* and to allow his views to be published only after his death.

In view of these conditions Steno's *Prodromus* is remarkable for its generally untrammelled reasoning, although the concluding pages of the essay are given over to a somewhat labored effort to prove that his views are not incompatible with Scripture, and that the written word has supplemented his observation. It seems doubtful, however, that he would have escaped persecution had he not been a devout Catholic and, moreover, under the protection of a powerful prince, the Grand Duke Ferdinand II. Some indication of the atmosphere of Florence in Steno's time may be gained from Vincentius Viviani's certificate appended to the *Prodromus* and approving it for publication, "since I have recognized in it a perfectly sincere manifestation of the Catholic faith and of good morals, as in the very candid author, I have thought the same worthy of being entrusted to type" (p. 271).

Steno is the pioneer of the observational methods which dominate in modern science, but he was destined to pass away and be almost forgotten before the methods which he used were to be adopted by students of science. If we except Leonardo da Vinci, who like Steno was a Florentine by adoption and who antedated him by a century and a half, there was no writer upon natural science before the eighteenth century that in accuracy of observation, in cogency of reasoning, or in discrimination of judgment might be compared with the "learned Dane." In some measure Steno reflected, of course, the crude notions of his time. Thus we find him adopting, though apparently with some reserve, the doctrine of the four elements, fire, earth, air, and water. In the main, however, if we exclude the prolix introduction addressed "to the Most Serene Grand Duke" and the weak conclusion intended to prove the orthodoxy of his position, the *Prodromus* with but moderate changes may be made to harmonize with the science of the twentieth century. We must attribute it largely to the closeness of his observation of Nature and to his discriminating judgment, that Steno was not lured into wild speculations, as were so many in his time. One of his statements might well be printed in large letters and placed upon the walls of our laboratories and lecture rooms, as a warning to those who undertake scientific investigation. "The nurse of doubts," says Steno, "seems to me to be the fact that in the consideration of questions relating to nature those points which cannot be definitely determined, are not distinguished from those which can be settled with certainty" (p. 213). How much trouble would be saved if to-day scholars had this point oftener in mind!

The form of Steno's essay is geometrical, and this is responsible for the almost unintelligible title and the correlation of subjects which, interpreted in the elaborate differentiation of twentieth century science, appear somewhat incongruous. As stated in the introduction, the *Prodromus* is divided into four parts. The first of these contains among other things an inquiry into the origin of fossils. The second part is stated to be: "Given a substance possessed of a certain figure, and produced according to the laws of nature, to find in the substance itself evidences disclosing the place and manner of its production." In like geometrical form the third part discusses solids which are contained within solids. The concluding portion of the essay is largely a consideration of the prehistoric geological changes which Steno was able to read in the rocks of Tuscany.

The broad outlines of the Cartesian conception of matter were adopted by Steno, who regarded a natural body as an aggregate of imperceptible particles subject to the action of forces such as proceed from a magnet, from fire, or sometimes from light. A fluid differed from a solid in having its particles in constant motion and withdrawing from their neighbors, that is to say, changing their relative positions.

Some of Steno's greatest contributions to science lie in the field of crystallography, for he studied the growth of crystals and showed that those formed in the mountains must have developed in the same manner as crystals of niter separating from solutions in water. These grow, he said, by accretions of substance upon the surface of the crystal nucleus, and not as do plants and animals.

The prevalent columnar form of crystals and the variation of their habit through the occurrence of faces of variable size, Steno explained by the addition of substance on certain sides only of the growing crystal. The force which draws the substance out of the surrounding fluid he recognized to be inherent in the crystal itself, and this crystallizing force he happily likened to what we should to-day call the lines of force about a magnet.

It is hardly to be expected that, great as Steno was, he should in his day have discovered the important fact of the orientation of the molecules of crystals, but he did point to the striking peculiarity of light refraction that distinguishes the crystal from amorphous substances, such as glass. Steno was, however, the discoverer of the fundamental law of crystallography known as the *law of constancy of interfacial angles*. As usually stated, this law affirms that *no matter how much the faces of a crystal may vary in their size or shape, the interfacial angles remain constant, provided they are measured at the same temperature.* The absolutely empirical verification of this law was delayed until the invention of the reflecting goniometer in 1805. Barring the refinement of temperature variations, it was amply verified by Rome de l'Isle with the simple goniometer which he invented in 1783. It is clear, however, that Steno more than a century earlier fully grasped the principle of the law, and gave it some sort of crude experimental verification. In the explanation of his figures, Steno says (p. 272):

"Figures 5 and 6 belong to the class of those crystals which I could present in countless numbers to prove that in the plane of the axis both the number and the length of the sides are changed in various ways without changing the angles."

As a corollary to his deductions concerning the growth of crystals, Steno showed that so-called "phantom crystals" are no product of the action of the larger crystal, but existed first and were enveloped through continued growth of the crystal nucleus.

In the realm of geology we owe to Steno the first clear enunciation of some of those great principles which to-day we assume to be axiomatic only because so much has been built upon them as a foundation. That rocks in the main result from sedimentation in water is thus expressed in the *Prodromus* (p. 219):

"The strata of the earth, as regards the manner and place of production, agree with those strata which turbid water deposits."

The reasons for this belief are most cogent: "The strata of the earth are due to the deposits of a fluid, (1) because the comminuted matter of the strata could not have been reduced to that form unless, having been mixed with some fluid and then falling from its own weight, it had been spread out by the movement of the same superincumbent fluid; (2) because the larger bodies contained in these same strata obey, for the most part, the laws of gravity, not only with respect to the position of any substance by itself, but also with respect to the relative position of different bodies to each other" (p. 227).

It is further clearly shown how marine deposits may be distinguished by their character from those deposits which are laid down in fresh water upon the continents, as well as from the ejectamenta of volcanoes. The origin of variation in the character of strata from place to place, and of the alternation of layers of different characters, are all discussed with a clear understanding of the actual conditions. The great principle that the order of superposition of beds determines the age of formations, is given its first expression (p. 230):

"At the time when any given stratum was being formed, all the matter resting upon it was fluid, and, therefore, at the time the lowest stratum was being formed, none of the upper strata existed."

Likewise it is pointed out that sedimentary formations were either laid down in definite basins of deposition or were universal in their extent. The original horizontality of sedimentary formations is now regarded as one of the great fundamental principles of geology. Steno says of the strata "that the upper surface was parallel to the horizon, so far as possible; and that all strata, therefore, except the lowest, were bounded by two planes parallel to the horizon. Hence it follows that strata either perpendicular to the horizon or inclined toward it, were at another time parallel to the horizon" (p. 230).

If strata are no longer in a horizontal position, it indicates, says Steno, subsequent disturbance of them; and this may be due either to uplift "by violent thrusting up of the strata," or "spontaneous slipping or downfall of the upper strata after they have begun to form cracks, in consequence of the withdrawal of the underlying substance, or foundation" (p. 231).

These changes in position of the strata are according to Steno the chief cause of mountains, and he pretty clearly distinguishes three of the more important mountain types; namely, (1) block or fault mountains, (2) volcanic mountains, and (3) mountains of erosion. The relation of earthquakes to the formation of mountains is indicated with a much nearer approach to present beliefs than is to be found in any save Robert Hooke and comparatively recent writers.

The fissures which form in the strata were recognized by Steno to be the passageways or channels for the movement of underground water, and for subterranean gases as well. These crevices are thus the places where veins of mineral are formed. The storehouses of the precious metals being brought about by natural processes, the foolishness of those who employ the divining rod for the locating of them is pointed out. An imperfect notion of the manner of replacement of one mineral by another seems to have been gained by Steno from his studies.

In the description of the figures — a most important part of the essay — a clear conception is revealed of the relative order of age of strata, of the alternation of transgression and recession of the sea over the same places, and of the nature of a structural unconformity, whereby one set of strata comes to overlie another from which it differs in its lesser angle of inclination. Here Steno gives us the results of his careful field observations in the vicinity of Florence. His figures may, therefore, be regarded as the earliest geological sections ever prepared.

Over the origin of fossils war had long been waged in Steno's time. Like Leonardo, a century and a half before, Steno declared that fossils were petrified remains of plants and animals which had once existed.

Steno's activity in biological studies is brought out in his elaborate examination of the structure of the shells of mollusks. His description of the subdivisions of the shells and the division of these into filaments, and of the various surfaces formed by the aggregation of these filaments, is suggestive of the methods of modern histological science. He shows that the substance of the filaments is developed from a fluid exuded through the outer surface of the animal. The structure of pearls, and their relation to the growing mollusks, is discussed at considerable length.

In treating the length of geological time, Steno was clearly hampered by the church doctrine of the time, to which he himself subscribed. Accepting as correct the Usher chronology of the Scriptures and the Noachian conception of the universal deluge, it is small wonder that Steno fell into error in evaluating geological time. "There are those," he said, "to whom the great length of time seems to destroy the force of the remaining arguments, since the recollection of no age affirms that floods rose to the place where many marine objects are found to-day, if you exclude the universal deluge, four thousand years, more or less, before our time" (p. 258). He thinks it possible to affirm that the shells dug from the hill on which Volterra was built, were formed more than three thousand years ago. The remains of elephants and extinct animals which were found in the valley of the Arno, and which we now know crossed from Africa on a land bridge in Tertiary times, Steno was forced to regard as the mired pack animals which had been brought by Hannibal's army on its way to besiege Rome.

Now that the Latin text of Steno's work has become available through republication, it seems opportune to make his argument accessible in English, and it is believed that Dr. Winter's rendering of the learned Dane's *Prodromus*, with annotations and with a brief account of his life and writings, will be welcomed by students of natural science.

WM. HERBERT HOBBS.

ANN ARBOR, Michigan,
  February, 1916.

INTRODUCTION

I. THE LIFE OF NICOLAUS STENO

NICOLAUS STENO,[1] the son of a goldsmith, Steen Pedersen, was born in Copenhagen, January 10, 1638.[2]

'From early childhood,' he wrote in 1680,[3] 'the association with those of my own age had little charm for me. For I was constantly in poor health from my third to my sixth year, and was accordingly under the continual care of my parents and older friends. As a result, I grew to prefer the conversation of older people, especially when they spoke of religion, to the frivolous chatter of younger companions. In my journeys, also, I kept away, as much as possible, from idle and dangerous people and sought friendship with those who had won repute through their upright life or their learning.'

After acquiring a thorough training in ancient and modern languages and mathematics in the grammar school of his native city, Steno entered its University in 1656, where he took up the study of medicine and had among his instructors the distinguished scientists Thomas Bartholin, Borrichius (Ole Borch), and Simon

---

[1] Niels Steensen, the Danish form of the name, in accordance with the learned custom of his day was Latinized by its bearer as Nicolaus Stenonis. The current form, Steno, is due to the mistaken idea that Stenonis was a genitive case. The spelling in French is Sténon and in Italian Stenone. Cf. Vilhelm Maar, *Nicolai Stenonis Opera Philosophica* (2 vols., Copenhagen, 1910), Vol. I, p. 1, note 1. According to custom, Steno took his surname from his father's given name; see Plenkers, *Der Däne Niels Stensen, Ein Lebensbild* (Freiburg im Br., 1884), p. 3, note 1.

The sources of Steno's life, consisting chiefly of letters and unpublished manuscripts, are given by Plenkers, *op. cit.*, pp. v, vi. To Plenkers, Wichfeld (*Erindringer om den danske Videnskabsmand Niels Stensen* in *Historisk Tidsskrift*, 3 Raekke, 4 Bind, Kjøbenhavn, 1865, pp. 1–109) and Maar (Vol. I, pp. i–xi) I am chiefly indebted for the biographical material here given.

[2] The *Encyclopædia Britannica*, in its eleventh edition, presents an inadequate biography of Steno in seventeen lines, and incorrectly gives 1631 as the date of his birth. The ninth edition contains no biographical notice. The error appears also in the account by Chéreau in the *Dictionnaire Encyclopédique des Sciences Médicales* (Troisième Série, Tome Onzième, Paris, 1883, pp. 689–691), in which January 1, 1631, is given as the date.

[3] *Defensio et plenior elucidatio epistolae de propria conversione*, Hannover, pp. 18, 19; quoted by Plenkers, *Niels Stensen*, pp. 3, 4.

Paulli. In all probability he took no degree,[1] for the times were troublous, and Steno, like other students, helped defend Copenhagen during its siege by Carl Gustav, king of Sweden.

In 1660 Steno went to Amsterdam to continue his studies and was warmly received by Gerard Blaes, the anatomist, to whom he had been recommended by Bartholin. His stay of four months in Amsterdam was made memorable by the discovery on April 7, 1660, of the parotid duct, which is still known as the *ductus Stenonianus*.[2] Blaes, however, claimed the discovery for his own, and a warm controversy ensued, in which Steno defended himself with ability and dignity. On April 22, 1661, Steno wrote to Thomas Bartholin from Leyden:[3]

'Since you urge me in your letter to publish an account of the exterior salivary duct, I am constrained to explain to you briefly the envy which this bit of a discovery (*inventiuncula*) has won for me, and also the result of this envy; not with the purpose of seeking fame in trifles,[4] but in order to free myself from the hateful charge of stealing what does not belong to me. For I am sorry that the necessity has been laid upon me of being forced to say much upon a subject of no importance, or else to submit to the base brand of shame. A due consideration of the matter will show that it is not worth making much ado about. For a similar duct[5] had been previously discovered, and even the very duct in question had been observed by Casserius[6] although he called it a muscle. . . . Since, however, the charge imputed to me by reason of that duct does not

---

[1] Wichfeld, *op. cit.*, p. 6, says that Steno went to Amsterdam in 1660 as "Dr. physices." He is followed by de Angelis in *Biographie Universelle* (Michaud; *Nouvelle Édition, Tome Quarantième*, p. 209), by Chéreau in the *Dictionnaire Encyclopédique des Sciences Médicales* (p. 689), and by Hughes in *Nature* (Vol. 25, 1882, p. 484). Plenkers (*Niels Stensen*, p. 11, note 5) gives good evidence for believing that no degree had been conferred, and Maar (*Opera Philosophica*, Vol. I, p. ii) implies as much.

[2] It appears that the parotid duct was observed independently by Needham in 1655, but his results were not published until 1667 (Maar, *op. cit.*, Vol. I, p. iii). Steno's treatise bears the following title and date: *De Glandulis Oris et Novis Inde Prodeuntibus Salivae Vasis*, Lugd. Batav. Anno 1661. It is printed by Maar, *op. cit.*, Vol. I, pp. 9–51.

[3] *De Prima Ductus Salivalis Exterioris Inventione et Bilsianis Experimentis*, Lugd. Batav. Ao (anno) 1661, 22 ap. (Aprilis). Printed by Maar, *op. cit.*, Vol. I, pp. 1–7.

[4] *In mustaceo laureolam quaeram* means literally 'look for a laurel-wreath in a cake.' Cicero uses the proverb in writing to his friend Atticus, V. 20, 4.

[5] *Ductus Whartonianus*, for which see *Adenographia . . . Auctore Thoma Whartono*, London, 1656, c. XXI, p. 129; Maar, *op. cit.*, Vol. I, p. 222.

[6] *De Vocis Auditusque Organis Historia Anatomica*, Ferrara, 1600, tab. V, p. 27, d, according to Maar, *op. cit.*, Vol. I, p. 222.

permit me to keep silent, I shall tell you the entire affair, as pupil to preceptor, and shall leave the decision to your judgment. ...

'It is a year now since I was hospitably received by Blaes. After waiting three weeks for a chance to secure anatomical material, I asked the distinguished man whether I might be permitted to dissect with my own hand such material as I could buy for myself. He gave his consent, and fortune so favored me that in the first sheep's head, which I had bought on April 7 and was dissecting alone in my room, I found a duct which, so far as I knew, had been described by no one before. I had removed the skin and was preparing to dissect the brain when I decided to examine first the ducts. With this end in view I was exploring the courses of the veins and arteries when I noticed that the point of my knife was no longer closely confined between tissues but moved freely in a large cavity, and presently I heard the teeth themselves resound, as I thrust my knife forward.

'In amazement at the discovery I called in my host (Blaes) that I might hear his opinion. First he ascribed the sound to the violence of my thrust, then resorted to calling it a freak of Nature, and finally referred to Wharton.[1] But inasmuch as that did not help, and the ducts, which had been handled carelessly, allowed no further investigation, I decided to examine them another time more carefully. I succeeded, although not so well, a few days later with a dog's head. Since its affinity to the inferior duct indicated the function of the one I had found, I told Jacob Henry Paulli, my intimate friend, that I had discovered a salivary duct, and I added a description of it. But since I knew that something like it had been discovered before and could not determine whether this identical duct had been examined, I remained silent until I could find opportunity to consult Sylvius about it. After he had heard my account he determined to seek the duct in man, and having found it he demonstrated it to spectators on several occasions.'[2]

Steno then proceeds to show that Blaes's brother, who was in Amsterdam at the time and was thoroughly conversant with the discovery, had accredited it to Steno in letters to Eysson, Professor

---

[1] See p. 176, note 5.

[2] Maar, *op. cit.*, Vol. I, pp. 3–5. For a convenient description of the duct and the relation of Steno's work to that of Richard Hale see de Angelis, *Biographie Universelle* (Michaud), *Nouvelle Édition, Tome Quarantième*, p. 209.

at Groningen. And furthermore, while Blaes mentioned the duct in his *Medicina Generalis*, which appeared in 1661, a year after the discovery, he could account for neither the beginning nor the end of the duct.[1]

Meanwhile Steno had gone to Leyden, where he remained from 1660–1664. Here he worked under van Horne, the surgeon, and Franciscus de la Boë Sylvius, the distinguished anatomist who discovered the Sylvian aqueduct. Among his intimate friends were men of widely differing attainments. The brilliant young Dane seems, in fact, to have had a genius for friendship. No fellowship could fail to stimulate reflection which included such men as Jan Swammerdam, the naturalist, whom Steno had previously known at Amsterdam; Borrichius, his old teacher, who had come from Copenhagen; Matthias Jacobaeus, Professor at Copenhagen, and later Bishop of Aarhus in Jutland; Peter Schumacher, who later became Count Griffenfeldt and High Chancellor of Denmark; Jacob Golias, Professor of Arabic at Leyden, and Baruch Spinoza, the philosopher, who was then living at Rijnsburg, a suburb of Leyden. But great as the influence of these men was, it was, perhaps, less telling for his subsequent spiritual development than the religious tolerance which Holland alone of European countries then afforded.[2]

While pursuing his anatomical studies in Leyden, Steno learned of the death of his step-father,[3] and thereupon returned to his native city. Disappointed in his expectation of gaining a professorship, Steno set out in the same year, 1664, for Paris, where he and Swammerdam lived with the naturalist Thévenot. It was in the latter's house that Steno delivered his discourse on the anatomy of the brain.[4] This treatise shares with the *Prodromus* the virtues of

[1] Maar, *op. cit.*, Vol. I, p. 223; Plenkers, *Niels Stensen*, pp. 12-14, and Wichfeld, *Erindringer om Niels Stensen*, pp. 7, 8.

[2] Maar, *Opera Philosophica*, Vol. I, pp. iv-v.

[3] Steno's father died in 1644, and his mother, Anna Nilsdatter, had contracted a second marriage with Johannes Stichman. Her death followed closely upon that of the latter. See Plenkers, *Niels Stensen*, pp. 3, 22, 25.

[4] *Discours sur l'anatomie du cerveau*, first printed in Paris in 1669; it is reprinted by Maar, *op. cit.*, Vol. II, pp. 3-35.

Jacques Bénigne Winslow, a grand-nephew of Steno, and himself a scientist of note, was so impressed by the treatise that he printed it in full in his *Exposition Anatomique* (Paris, 1732), pp. 641-659. The preface of the work closes with this remarkable acknowledgment:

"Je finis en avertissant avec une sincere reconnoissance, que le seul Discours de feu M. Stenon sur l'Anatomie du Cerveau, a été la source primitive et le modele general de toute

lucidity and scientific objectivity. At a time when fantastic metaphysics were rife, Steno trusted only to induction based upon experiment and observation. Combating the theories of Descartes and Willis in particular, the author prefaces his discourse with the candid admission:

"Au lieu de vous promettre de contenter vostre curiosité, touchant l'Anatomie du Cerveau; ie vous fais icy une confession sincere et publique, que ie n'y connois rien. Ie souhaiterois de tout mon cœur, d'estre le seul qui fust obligé à parler de la sorte; car ie pourrois profiter auec le temps de la connoissance des autres, et ce seroit un grand bon-heur pour le genre humain, si cette partie, qui est la plus delicate de toutes, et qui est sujette à des maladies tres-frequentes, et tres-dangereuses, estoit aussi bien connuë, que beaucoup de Philosophes et d'Anatomistes se l'imaginent. Il y en a peu qui imitent l'ingenuité de Monsieur Sylvius, qui n'en parle qu'en doutant, quoy qu'il y ait travaillé plus que personne que ie connoisse. Le nombre de ceux à qui rien ne donne de la peine, est infailliblement le plus grand. Ces gens qui ont l'affirmative si prompte, vous donneront l'histoire du cerveau, et la disposition de ses parties, avec la mesme asseurance, que s'ils avoient esté presens à la composition de cette merveilleuse machine, et que s'ils avoient penetré dans tous les desseins de son grand Architecte."[1]

Paris, however, did not hold Steno long. In the summer of 1665 we find him in Florence,[2] where he was soon attached to the court of the Grand Duke Ferdinand II. Upon the recommendation of Thévenot and Viviani, Steno was appointed physician to the Grand Duke, with a house and pension. He was also given a position in the hospital of Santa Maria Nuova. This patronage, gratefully acknowledged in the present treatise, and the opportunity for travel

ma conduite dans les travaux Anatomiques. Je l'ai inseré dans le Traité de la Tête, croyant faire plaisir au Public de lui communiquer de nouveau cette Piece, qui étoit devenue rare, et qui renferme beaucoup d'excellens avis, tant pour éviter le faux et l'imaginaire, que pour découvrir le vrai et le réel, non seulement par rapport à la structure et aux usages des parties, mais aussi par rapport à la maniere de faire les Dissections et les Figures Anatomiques."

The first and only complete edition of *L'Autobiographie de Jacques Bénigne Winslow* is that of Maar, Copenhagen, 1912. I am indebted to it for the foregoing passage, p. xxiv.

[1] Maar, *Opera Philosophica*, Vol. II, p. 3.

[2] I have followed Maar, *Opera Philosophica*, Vol. I, p. vi. Plenkers, *Niels Stensen*, p. 30, and Wichfeld, *Erindringer om Niels Stensen*, p. 17, give 1666 as the date of Steno's arrival in Florence.

which his position at court afforded him, made possible Steno's scientific researches.

Now followed the happiest and most productive period of Steno's life. Ferdinand II, although a weak prince, was a generous patron of art and science. The Accademia del Cimento, founded in 1657 by Leopold de' Medici, the brother of Ferdinand, was the center of a learned group including Vincenzo Viviani, the pupil and biographer of Galileo, Francesco Redi, poet and naturalist, Carlo Dati, the scientist, and Lorenzo Magalotti, the versatile secretary of the society. 'I have the honor,' Redi wrote to Athanasius Kircher,[1] 'to serve at a court where distinguished men gather from all parts of the world. In their wanderings they bring and seek in exchange the fruits of high endeavor, and so warm is their welcome that they fancy themselves transported to the mythical gardens of the Odyssey.'

Through the influence of Maria Flavia del Nero, a nun who had long been in charge of the apothecary connected with Santa Maria Nuova, also Lavinia Felice Cenanni Arnolfini, the wife of the ambassador from Lucca, and Emilio Savignani, a Jesuit priest, Steno was induced to embrace Catholicism. He was deeply religious by nature, and there can be no question about the sincerity of his conversion. His seriousness as a lad, and the impression made upon him by the religious tolerance in Holland, have been mentioned. Furthermore, since meeting the eloquent Bossuet in Paris he had been pondering deeply the question of Catholicism versus Protestantism.[2] He was finally received into the Church, December 8, 1667. Five days later Viviani wrote to Magalotti, who was then in Flanders:

'My very dear friend, N. Steno, who lacked only this to make him adorable, so to say, has turned back to life on the day of the dead,[3] in that he has confessed the Catholic faith. His decision to take the final step gave great joy to His Highness (Ferdinand II) and all his friends. On the day of the Immaculate Conception,

---

[1] Quoted by Plenkers, *Niels Stensen*, p. 31.

[2] The question of Steno's conversion is treated at length by Plenkers (*Niels Stensen*, pp. 36-50), who includes in his account many of the letters that passed between Steno and his friends. The reasons which induced Steno to take this step were set forth by him in *Epistola de propria conversione* (Florence, 1677) and *Defensio et plenior elucidatio epistolae de propria conversione* (Hannover, 1680).

[3] All Souls' Day, November 2, is *Giorno de'Morti* in Italian.

after he had finally declared his conversion before the Nuntius also, he received a letter from his King, which he called an invitation, with the command to return as soon as possible. An annual pension of four hundred scudi was promised him from the day of his departure. There are no further stipulations, and he can expect an increase of this amount. Still, he is unwilling to begin the journey until he has learned whether His Majesty will support him in this way in spite of his change of belief. Inasmuch as we cannot hope that this will be the case, we have the prospect of keeping him with us.'[1]

According to Blondel[2] Steno wrote to Frederik III informing him of his change of belief. While these negotiations were in progress Steno composed his *Prodromus*[3] (1668). The original plan of writing the treatise in Italian was given up in favor of Latin, and when the remarkable essay appeared it was under the title *Nicolai Stenonis De Solido Intra Solidum Naturaliter Contento Dissertationis Prodromus*. It was printed at Florence, in 1669, with the full sanction of the papal authorities, among whom were his influential friends, Redi and Viviani.

The *Prodromus* was intended as a preliminary statement of principles which the author expected to elaborate more fully in a later, comprehensive work, which is referred to throughout as the Dissertation. The larger work, however, never appeared, probably because

---

[1] Quoted by Plenkers, *Niels Stensen*, p. 51. This order of Frederik III, dated 19 October, 1667, is still preserved in Copenhagen. Compare *op. cit.*, note 1.

[2] *Les Vies des saints pour chaque jour de l'année*, Paris, 1722, p. 738.

[3] The use of the word *Prodromus* to designate a treatise preliminary to a larger work is not found in classical Latin. The *New Oxford Dictionary* amply illustrates its occurrence in English, but the examples are from works subsequent to Steno's time. Francis Bacon (1561–1626) employs the word in the *Instauratio Magna, Prodromi sive Anticipationes Philosophiae Secundae* (edition of Spedding, Ellis, and Heath, Vol. V, p. 182). Larousse, *s.v. Prodrome*, not only gives the best definition of the word as used by Steno, but also cites an excellent example of a writer whose accomplishment, like Steno's, fell short of his original expectation:

"Ce mot a été employé pour désigner une préface, une introduction, un discourse préliminaire; mais, dans sa signification la plus généralement acceptée, il est le titre même d'un ouvrage destiné a préparer d'autres écrits dont il donne l'idée et auxquels il prépare le lecture. Il a été fait des livres de ce genre sur les matières théologiques et philosophiques. Il en existe aussi qui sont relatifs aux sciences exactes et naturelles. L'un des plus remarquables est celui que Candolle a publié sous ce titre: Prodrome du système du regne végétal (Paris, 1824 et suiv. in 8vo). Ce célèbre botaniste avait d'abord conçu le plan d'un ouvrage extrêmement vaste, qu'il intitula: Système naturel du règne végétal, et dont il fit paraître deux volumes (1818–1821, in 8vo); mais, comprenant que la vie d'un homme ne se suffirait pas à remplir ce plan, il y renonça et fit son *Prodrome*, recueil déjà fort vaste, présentant le répertoire des ordres, des genres, des espèces du règne végétal, et qu'il ne put terminer."

Steno's interest in geology had meanwhile given way to his interest in theology.[1] Brief as it is, the *Prodromus* remains one of the most noteworthy contributions to the science of geology, and especially the geology of Italy. Steno's habits of observation, analysis, and induction resulted in an enlightened exposition of geology considered from the petrological, palæontological, and stratigraphical point of view, at a time when many of his contemporaries were still satisfied with some of the absurdities of metaphysical speculation.[2] Steno's work, von Zittel remarks,[3] "already contained the kernel of much that has been under constant discussion during the two centuries which have passed since his death; and if one reads the most recent text-books of geology, it will be evident that science has not yet securely ascertained the share that is to be assigned to subsidence, to upheaval, to erosion, and to volcanic action in the history of the earth's surface conformation in different regions."

The journey to Denmark was not undertaken until a year after Steno's conversion, and then by a circuitous route. After visiting Rome, Naples, and Murano, he reached Innsbruck in May, 1669, Vienna and Prague in the late summer of the same year, and finally Amsterdam in the spring of 1670. Meanwhile Frederik III had died February 2, 1670, and Steno remained in Holland. Upon learning of the serious illness of his patron Ferdinand II he departed at once for Florence. When he arrived (1670), Cosimo III had already succeeded his father as Grand Duke of Tuscany. But the change in rulers brought no change in the warmth of Steno's welcome. Under Cosimo he arranged the minerals in the Pitti Palace, and continued his studies in geology.

In the Pitti Palace is a series of portraits of distinguished men who were associated with the Court of Ferdinand II and Cosimo III.

[1] The *Prodromus* was Steno's last scientific work of note. After his conversion (cf. p. 180) his interest in science rapidly waned. Leibnitz, who came to know and esteem Steno later in Hannover, in letters to Conring expresses deep regret that Steno had abandoned his earlier studies. See Gerhardt, *Die philosophischen Schriften von G. W. Leibniz* (Vol. I, Berlin, 1875), p. 185, and especially p. 193: "*Stenonium Episcopum doleo nunc a physiologicis studiis averti ad theologica vel ideo quia in his facilius quam in illis habebit parem.*"

[2] A striking instance of this is Kircher's *Mundus Subterraneus*, Amsterdam, 1665. Compare Maar, *Om Faste Legemer*, Copenhagen, 1902, p. ii ff.

[3] *History of Geology and Palæontology*, Eng. trans. (London, 1901), p. 27. Compare Huxley, *Nature*, Vol. 24 (1881), p. 453; A. von Humboldt, *Essai Géognostique sur le Gisement des Roches dans les deux Hémispheres* (Paris, 1823), p. 38; *Cosmos*, Eng. trans. (London, 1852), Vol. 2, pp. 347–348; M. J. P. Flourens, *De la Longévité humaine et de la Quantité de Vie sur le Globe* (Paris, 1855), pp. 211–215.

Among these is a portrait of Steno, by an unknown painter, evidently made in the period when the *Prodromus* was composed; this is reproduced in our Plate V.

On the recommendation of Count Griffenfeldt,[1] Christian V invited Steno to the Professorship of Anatomy in Copenhagen. The royal order, dated February 13, 1672, is still preserved in Copenhagen, and reads:

'Know that by special royal grace and favor We have allowed you, until further gracious increase, four hundred reichsthaler a year. This pension shall date from the time of your arrival here. For it is our gracious command and will that you undertake at once your journey to our Kingdom of Denmark, in order to be here as soon as possible. You will comply in humble obedience.'[2]

Steno's reply to Count Griffenfeldt is dated April 26:

'I thank your Excellency most humbly for your good will to me and wish with all my heart that God may grant me to prove to you, one day, my gratitude and willing service. I suppose that your Excellency already knows the reason why my answer is so late in arriving. For the letters containing the orders of His Royal Majesty did not reach Holland until April 3. From there they were forwarded to me yesterday, the twenty-fifth of April. I humbly ask you therefore to excuse my delay to our most gracious Lord and King. This morning I went to the Grand Duke to tell him of the command of His Royal Majesty, and I hope to receive leave for my final departure within a few days.'[3]

Steno arrived in Copenhagen July 3, 1672. His notable address on the reopening of the Theatrum Anatomicum[4] was as much a valedictory as an inaugural, for it marks the close of his scientific career. He soon became involved in religious controversies which made his tenure so disagreeable that he yearned for the old life at Florence. Accordingly, he resigned his position in the summer of 1674, and set out for Florence, visiting the Catholic Duke of Hannover, Johann Friedrich, on the way. Thence he came back to Amsterdam before proceeding to Italy. Upon his arrival in Florence, late in the year 1674, he was appointed tutor to the son of Cosimo III, and thenceforth gave up natural science, for which his keen powers

---

[1] See p. 178. [2] Quoted by Plenkers, *Niels Stensen*, p. 91.
[3] Plenkers, *op. cit.*, p. 91.
[4] Printed by Maar, *Opera Philosophica*, Vol. II, pp. 249-256.

of observation and analysis so admirably fitted him, in order to devote himself to questions of education and theology.

In 1675 Steno took Holy Orders; in the following year (September 14, 1676), Pope Innocent XI rewarded his zeal in attempting to convert his former friends[1] and co-religionists by appointing him Bishop of Titopolis,[2] *in partibus infidelium*, and Apostolic Vicar of Northern Germany and Scandinavia. In consequence, toward the end of 1677, he took up his residence in Hannover. An account preserved by Manni[3] gives a graphic picture of the austere life Steno's devotion now induced him to lead:

'The prelate lived and dressed as though he were the poorest person in the world. His position could only be inferred from his ecclesiastical garb, and even this was only serge. For he would not take the robes of his predecessor although they were offered to him at a low price. And notwithstanding the Duke made him an ample allowance to enable him to live as became his rank, he gave everything to the poor. For them he sacrificed everything. And he did this as long as we knew him. He even gave the gold necklace with a medallion containing a portrait of the Duke—he had received it on his second return from Denmark to Rome by way of Hannover[4]—to a friend with the injunction that it be bestowed upon the poor. When he had nothing else he sold his silver crucifix and costly bishop's ring to relieve the distress of others.'

Upon the death of Johann Friedrich, in 1679, and the accession of his Protestant brother, Duke Ernst August, Steno was forced to withdraw to Munster. Here he was appointed Suffragan Bishop to Ferdinand, Baron von Fürstenberg, the Bishop of Munster (1680). The latter died in 1683, and was succeeded by Archbishop Maximilian Heinrich. Steno had opposed his election and refused to celebrate mass in honor of the event. He therefore withdrew to Hamburg, where his self-imposed poverty and his asceticism alienated the Catholics themselves. They threatened to cut off his nose and ears, to drive him from the city, and even to kill him. In his

---

[1] Plenkers, *Niels Stensen* (pp. 122, 123), quotes, among other letters, an interesting appeal to Spinoza. The latter did not reply.

[2] An old bishopric in Isauria.

[3] *Vita del letteratissimo Mons. N. Stenone* (Florence, 1775), p. 229; quoted by Plenkers, *Niels Stensen*, p. 131.

[4] This statement is inexact; Steno did not go to Denmark in 1670, and in 1674 his objective was Florence, not Rome.

PLATE VI.

PORTRAIT OF STENO AS VICAR OF SCHWERIN.

trouble he began to long for the peace and friendships of Italy,[1] and was preparing to return when the missionary post at Schwerin was offered to him. He accepted it in 1685 as a call to further service. But the change meant only increased fasting and abject poverty, to which he succumbed November 26, 1686. At the request of Cosimo III Steno's body was taken to Florence and laid in the famous San Lorenzo.

The physical change which Steno's self-denial entailed is strikingly shown in his portrait as Vicar of Schwerin. The original, by an unknown artist, is still in Schwerin. Until recently it was his only known portrait. An excellent copy, reproduced in our Plate VI, is in possession of the Anatomical Institute of the University of Copenhagen.

On the walls of the cloister of San Lorenzo there is to-day a medallion portrait of Steno, surrounded by a marble wreath, with the following inscription, in black letters, beneath it:

NICOLAI · STENONIS · IMAGINEM · VIDES · HOSPES
QVAM · AERE · COLLATO · DOCTI · AMPLIVS · MILLE
EX · UNIVERSO · TERRARUM · ORBE · INSCULPENDAM
CURARUNT · IN · MEMORIAM · EJUS · DIEI · IV · CAL · OC-
TOBR · AN · M · D · CCC · LXXXI · QUO · GEOLOGI · POST · CON
VENTUM · BONONIAE · HABITUM · PRAESIDE · JOANNE
CAPELLINIO · EQUITE · HUC · PEREGRINATI · SUNT · AT-
QUE · ADSTANTIBUS · LEGATIS · FLOR · MUNICIPII · ET
R · INSTITUTI · ALTIORUM · DOCTRINARUM · CINERES
VIRI · INTER · GEOLOGOS · ET · ANATOMICOS · PRAE-
STANTISSIMI · IN · HUJUS · TEMPLI · HYPOGEO · LAUREA
CORONA · HONORIS · GRATIQUE · ANIMI · ERGO · HONE-
STAVERUNT

The medallion portrait is by Vincenzo Consani. Plenkers (*Niels Stensen*, p. 88) quotes the inscription, but does not divide it properly into lines. My own transcript was made in Florence June 20, 1911. The Latinity of the inscription is open to criticism; ALTIORUM in line 9 should be ALTIARUM, and the hyphens at the ends of lines 4, 7, 10, 12 are not in accordance with ancient usage. I add a translation:

[1] Indicated in the letters to Madame Arnolfini (Plenkers, *Niels Stensen*, p. 178).

186   INTRODUCTION

'Friend, you behold the likeness of Nicolaus Steno. To it more than a thousand men of learning, from all parts of the world, contributed. They made provision for the carving of it in memory of this day, the twenty-eighth of September, in the year 1881, when the Geologists, after the Congress at Bologna, under the Presidency of Cavaliere Giovanni Capellini, journeyed hither, and in the presence of delegates representing the City of Florence and the Royal Institute of Higher Studies, in the cloister of this church, as a testimonial of respect and gratitude honored with a laurel crown a man of surpassing distinction among Geologists and Anatomists.'

The official account of the events recorded in the inscription is in itself a commentary of sufficient interest to warrant reprinting here, particularly because of its estimate of the value of Steno's contribution to geology:[1]

Ils allèrent en suite rendre hommage aux restes de Sténon qui reposent dans une tombe de plus modestes, dans la crypte souterraine de la chapelle des Médicis, à San Lorenzo. Les chanoines de la Basilique se tenaient, pour les recevoir, au pied de l'escalier qui descend dans la crypte.

Là, M. le président Capellini invita à prendre la parole l'éminent représentant des études d'archéologie préhistorique, M. Waldmar Schmidt, de Copenhague. Notre savant confrère s'exprima en ces termes:

"Messieurs, Au moment où les membres du second Congrès géologique international sont réunis dans la célèbre église de San Lorenzo, devant la tombe de Nicolas Sténon, vous permettrez, je l'espère, au seul représentant du pays où est né Sténon, d'exprimer au noms de ses compatriotes les plus chaleureux remercîments à la ville de Florence pour l'hommage qu'elle a rendu à leur concitoyen.

"Comme vous le savez, à une époque où les sciences naturelles n'étaient pas encore sorties de leur première enfance, Nicolas Sténon a jeté les fondements de la géologie; et par ses études, par ses observations, par son génie perspicace, il est arrivé à énoncer, sur divers points de la science, des vues dont les géologues de notre siècle, après tant de nouvelles recherches, ont reconnu l'exactitude.

"Sténon était né en Danemark et c'est là qu'il fit ses premières

---

[1] *Congrès Géologique International* ... *Compte Rendu de la 2me Session, Bologne*, 1881, pp. 249–251. See also the brief account in *Bolletino del R. Comitato Geologico d' Italia*, vol. 12 (1881), pp. 379, 380.

études. Mais c'est en Italie qu'il a accompli ses merveilleuses découvertes et posé les bases de la géologie. Il y fut reçu avec cette splendide hospitalité qui nous a nous mêmes accueillis partout d'abord à Bologne, aujourd'hui à Florence.

"L'Italie fut sa seconde patrie, et ses restes mortels reposent dans ce temple magnifique, dans lequel on admire les œuvres des plus grands artistes du monde.

"Quand Sténon abandonna le Danemark pour venir se fixer dans ce beau pays, la science géologique ne déserta pas avec lui la patrie scandinave.

"Ne dois-je pas, Messieurs, vous rappeler à cette occasion que si le Danemark a eu Sténon, un autre pays scandinave, la Suède, a eu Linné. Comme l'un avait établi les fondements de la stratigraphie, l'autre posa les bases de la géologie physique. . . . Vous me permettrez donc, Messieurs, de joindre à mes remercîments pour la ville de Florence et son syndic qui nous ont fait un si magnifique accueil, l'expression de ma reconnaissance pour celui qui par un beau travail a fait connaître, je ne dirai pas le nom de Sténon qui était déjà assez connu, mais sa vie, son origine et son pays natal: toute ma gratitude à M. Capellini, auteur de la Vie de Sténon et président du deuxième Congrès international de géologie à Bologne."

Il (Capellini) ajoute que son but était de faire mieux connaître ce grand homme dont le souvenir doit être sacré pour tous les géologues, et sur la tombe duquel il est heureux de tendre la main à M. W. Schmidt, afin de reserrer entre l'Italie et le Danemark les liens d'affection que rappelle cette illustrée mémoire. . . .

Le soir, le cercle philologique, le cercle des ingénieurs et le club alpin ouvrirent gracieusement leur salles aux congressistes. Mais avant de se rendre à ces amiables invitations, ils furent conviés à un dîner à l'hôtel Minerva par le président Capellini, et, après, une souscription fut ouverte par ses soins pour placer sur le tombe de Sténon une pierre dont l'inscription rappellerait à la fois les glorieux titres scientifiques du célèbre Danois, et la visite faite à sa tombe par les membres du Congrès géologique international. . . .

II. THE WRITINGS OF STENO

Steno's published works may be grouped under three heads: Anatomy, Geology, and Theology. The scientific treatises have all been reprinted by Maar, *Opera Philosophica* (2 vols., Copenhagen, 1910); no complete edition of his many interesting letters, and of his theological writings, has yet appeared. The following list, compiled from Maar and Plenkers, gives the full title of each published work, its date, and place of original publication. In the case of the scientific treatises references are given also to Maar's edition.

1. ANATOMY

1. *a. Disputatio Anatomica de Glandulis Oris et Nuper Observatis inde Prodeuntibus Vasis Prima.* Leyden, 1661 (July 6).

   b. Disputatio . . . Secunda. Leyden, 1661 (July 9).

   These two articles appeared together in:

   De Glandulis Oris et Novis earundem Vasis Observationes Anatomicae. Leyden, 1661. Printed by Maar, Vol. I, pp. 9-51 (Number II).

2. *Observationes Anatomicae, Quibus Varia Oris, Oculorum, et Narium Vasa Describuntur, Novique Salivae, Lacrymarum et Muci Fontes Deteguntur, et Novum Nobilissimi Bilsii de Lymphae Motu et Usu Commentum Examinatur et Rejicitur.* Leyden, 1662.

   This volume includes four treatises:

   a. De Glandulis Oris, etc., pp. 1-54; a reprint of the two Disputationes. Maar, Vol. I, pp. 9-51 (No. II).

   b. Responsio ad Vindicias Hepatis Redivivi, Qua Tela, Quae in Praesidem Celeberr. Dn. Johannem van Horne direxerat Clar. Antonius Deusingius, a Thesium Authore Excipiuntur, et Evanida Ostenduntur. Pp. 55-78.

   This treatise bears the date 28 November, 1661. Maar, Vol. I, pp. 59-73 (No. IV).

c. De Glandulis Oculorum Novisque earundem Vasis Observationes Anatomicae, Quibus Veri Lacrymarum Fontes Deteguntur. Pp. 79–100. Maar, Vol. I, pp. 75–90 (No. V).

d. Appendix de Narium Vasis. Pp. 101–108. Maar, Vol. I, pp. 91–97 (No. VI).

3. *Apologiae Prodromus, Quo Demonstratur, Judicem Blasianum et Rei Anatomicae Imperitum Esse, et Affectuum Suorum Servum.* Leyden, 1663. Maar, Vol. I, pp. 143–154 (No. XIII).

4. *De Musculis et Glandulis Observationum Specimen Cum Epistolis Duabus Anatomicis.* Copenhagen, 1664. The *De Musculis*, etc., is printed by Maar, Vol. I, pp. 161–192 (No. XV).

The first of the two letters was written to Willem Piso, and is entitled *De Anatome Rajae Epistola.* Dated April 24, 1664, Copenhagen. Pp. 48–70. Maar, Vol. I, pp. 193–207 (No. XVI). The second was written to Paul Barbette, and is entitled *De Vitelli in Intestina Pulli Transitu Epistola.* Dated June 12, 1664, Copenhagen. Pp. 71–84. Maar, Vol. I, pp. 209–218 (No. XVII).

5. *De Prima Ductus Salivalis Exterioris Inventione, et Bilsianis Experimentis.* A letter to Thomas Bartholin, dated April 22, 1661, Leyden. First printed in Bartholin's *Epistolae Medicinae*, Cent. III, 1667, No. XXIV. Maar, Vol. I, pp. 1–7 (No. I).

6. *Variae in Oculis et Naso Observationes Novae.* A letter to Bartholin, dated September 12, 1661, Leyden. First printed in *Epist. Med.*, Cent. III, 1667, No. LVII. Maar, Vol. I, pp. 53–58 (No. III).

7. *Sudorum Origo ex Glandulis. De Insertione et Valvula Lactei Thoracici et Lymphaticorum.* A letter to Bartholin, dated January 9, 1662, Leyden. First printed in *Epist. Med.*, Cent. III, 1667, No. LXV. Maar, Vol. I, pp. 99–103 (No. VII).

8. *Cur Nicotinae Pulvis Oculos Clariores Reddat. De Lactea Gelatina Observatio.* A letter to Bartholin, dated May 21, 1662, Leyden. First printed in *Epist. Med.*, Cent. IV, 1667, No. I. Maar, Vol. I, pp. 105–111 (No. VIII).

9. *Observationes Anatomicae in Avibus et Cuniculis.* A letter to Bartholin, dated August 26, 1662, Leyden. First printed in *Epist.*

Med., Cent. IV, 1667, No. XXVI. Maar, Vol. I, pp. 113–120 (No. IX).

10. *De Vesiculis in Pulmone. Anatome Cuniculi Praegnantis. In Pulmonibus Experimenta. De Lacteis Mammarum. In Cygno Observationes.* A letter to Bartholin, dated March 5, 1663, Leyden. First printed in *Epist. Med.*, Cent. IV, 1667, No. LV. Maar, Vol. I, pp. 129–136 (No. XI).

11. *Nova Musculorum et Cordis Fabrica.* A letter to Bartholin, dated April, 1663, Leyden. First printed in *Epist. Med.*, Cent. IV, 1667, No. LXX. Maar, Vol. I, pp. 155–160 (No. XIV).

12. *Elementorum Myologiae Specimen, seu Musculi Descriptio Geometrica. Cui Accedunt Canis Carchariae Dissectum Caput, et Dissectus Piscis ex Canum Genere.* Florence, 1667.

The *Elementorum Myologiae Specimen* is printed by Maar, Vol. II, pp. 61–111 (No. XXII). The second treatise, entitled *Canis Carchariae*, etc., is printed by Maar, Vol. II, pp. 113–145 (No. XXIII). The third, *Dissectus Piscis*, etc., is printed by Maar, Vol. II, pp. 147–155 (No. XXIV).

13. *Discours sur l'Anatomie du Cerveau.* Paris, 1669. Maar, Vol. II, pp. 1–35 (No. XVIII).

14. *Figurae Explicatio. Receptaculi Sanguinis Circulus per Ventriculorum Cordis Separationem ab Invicem Manifestior Redditus.*

First printed in Bartholin's *Anatome, Quartum Renovata*, Leyden, 1673, pp. 805–807. Maar, Vol. II, pp. 279–282 (No. XXXIII).

15. *Embryo Monstro Affinis Parisiis Dissectus.* First printed in *Acta Hafniensia*, Vol. I, 1673, pp. 200–203. Maar, Vol. II, pp. 49–53 (No. XX).

16. *Uterus Leporis Proprium Foetum Resolventis.* In *Acta Hafn.*, Vol. I, 1673, pp. 203–207. Maar, Vol. II, pp. 55–60 (No. XXI).

17. *De Vitulo Hydrocephalo.* In *Acta Hafn.*, Vol. I, 1673, pp. 249–262. Maar, Vol. II, pp. 229–239 (No. XXVIII).

18. *In Ovo et Pullo Observationes.* In *Acta Hafn.*, Vol. II, 1675, pp. 81-92. Maar, Vol. II, pp. 37-47 (No. XIX).

19. *Ex Variorum Animalium Sectionibus hinc inde factis Excerptae Observationes circa Motum Cordis, Auricularumque et Venae Cavae.* In *Acta Hafn.*, Vol. II, 1675, pp. 141-147. Maar, Vol. I, pp. 121-127 (No. X).

20. *Observationes Anatomicae Spectantes Ova Viviparorum.* In *Acta Hafn.*, Vol. II, 1675, pp. 210-218. Maar, Vol. II, pp. 157-166 (No. XXV).

21. *Ova Viviparorum Spectantes Observationes factae Jussu Serenissimi Magni Ducis Hetruriae.* In *Acta Hafn.*, Vol. II, 1675, pp. 219-232. Maar, Vol. II, pp. 167-179 (No. XXVI).

22. *Lymphaticorum Varietas.* In *Acta Hafn.*, Vol. II, 1675, pp. 240-241. Maar, Vol. I, pp. 137-142 (No. XII).

23. *Historia Musculorum Aquilae.* In *Acta Hafn.*, Vol. II, 1675, pp. 320-345. Maar, Vol. II, pp. 257-277 (No. XXXII).

24. *Prooemium Demonstrationum Anatomicarum in Theatro Hafniensi Anni 1673.* In *Acta Hafn.*, Vol. II, 1675, pp. 359-366. Maar, Vol. II, pp. 249-256 (No. XXXI).

In addition to the foregoing treatises, Maar prints extracts from various sources, *op. cit.*, Vol. II, pp. 283-310 (Appendix, Nos. XXXIV, XXXV, XXXVI).

II. GEOLOGY

1. *De Solido intra Solidum Naturaliter Contento Dissertationis Prodromus.* Florence, 1669. Maar, Vol. II, pp. 181-227 (No. XXVII).

2. Letter to Cosimo III, in Italian, *On the Grotto above Gresta.* First printed by Fabroni, *Lettere Inedite di Uomini Illustri* (Florence, 1773-1775), Vol. II, no. 141, pp. 318-321. Maar, Vol. II, pp. 239-242 (No. XXIX).

3. Letter to Cosimo III, in Italian, *On the Grotto of Moncodine.* Dated August 19, 1671. First printed by Fabroni, *op. cit.*, no. 142, pp. 321-327. Maar, Vol. II, 243-248 (No. XXX).

III. THEOLOGY

1. *Ad Virum Eruditum cum Quo in Unitate S. R. E. desiderat Aeternam Amicitiam inire, Epistola detegens Illorum Artes Qui Suum de Interprete S. Scripturae Errorem S. Patrum Testimonio confirmare nituntur.* Florence, 1675.

2. *Epistola exponens Methodum Convincendi Acatholicum juxta D. Chrysostomum ex ejusdem Homilia XXXIII in Act. Apostolorum.* Florence, 1675.

3. *Epistola ad Novae Philosophiae Reformatorem de Vera Philosophia.* Florence, 1675.

4. *Epistola de Propria Conversione.* Florence, 1677.

5. *Scrutinium Reformatorum ad Demonstrandum Reformatores Morum fuisse a Deo, Reformatores autem Fidei et Doctrinae non fuisse.* Florence, 1677.

6. *Epistola de Philosophia Cartesiana.* Florence, 1677.

7. *Scrutinium Reformatorum d. i. Kurtzer Beweis dass Diejenigen Lehrer, so die Sitten der Menschen zu Verbessern Getrachtet, von Gott Gewesen, mit Nichten aber die Andern, so die Glaubenslehre zu Verbessern Gesuchet.* Hannover, 1678.

8. *Occasio Sermonum de Religione cum Jo. Sylvio.* Hannover, 1678.

9. *Examen Objectionis circa Diversas Scripturas Sacras et Earum Interpretationes Tamquam Divinas a Diversis Ecclesiis Propositas, D. Jo. Sylvio per Litteras a. 1670 Transmissum, modo Distinctius et Auctius in Lucem Editum, Ubi Omnes, Qui Reformatos Se Credunt, Nobis Nulla Unquam Fidei Reformatione Indigis Objiciunt, Se Solos Certos esse, Quod Deo Credant, Nostram autem Fidem Non Divina, Sed Humana Auctoritate niti.* Hannover, 1678.

10. *Tractatio de Purgatorio Cum Discursu utrum Pontificii an Protestantes in Religionis Negotio Conscientiae Suae Rectius Consulant.* Hannover, 1678.

11. *Katholische Glaubenslehre vom Fegfeur, mit Klaren Zeugnüssen aus dem H. Augustino Bewehret; nebenst Entdeckung Vier Grober Irrthümer des Dorschäi, indem Er Vorgibt dass Bellarminus das Fegfeur aus den H. H. Vättern Nicht Habe Erweisen Können*, etc. Hannover, 1678.

12. *Defensio et Plenior Elucidatio Scrutinii Reformatorum.* Hannover, 1679.

13. *Defensio et Plenior Elucidatio Epistolae de Propria Conversione.* Hannover, 1680.

14. *Parochorum Hoc Age, seu Evidens Demonstratio Quod Parochus Tenetur Omnes Alias Occupationes dimittere et Suae attendere Perfectioni ut Commissas Sibi Oves ad Statum Salutis Aeternae Ipsis a Christo Praeparatum Perducat.* Florence, 1683.

III. BIBLIOGRAPHY OF THE PRODROMUS

1. THE ORIGINAL EDITION

Nicolai Stenonis De Solido Intra Solidum Naturaliter Contento — Dissertationis Prodromus. Ad Serenissimum Ferdinandum II Magnum Etruriae Ducem. Florentiae. Ex Typographia sub signo Stellae MDCLXIX. Superiorum Permissu.[1]

The volume is a small quarto, the type page measuring seven and one sixteenth by four and three sixteenths inches, with wide margins. Including the margins, the page measures nine and one half by six and five sixteenths inches. Maar observes (*Opera Philosophica*, Vol. II, p. 355) that copies are extant without the wide margins, although he does not state where either these or the former may be found. The title-page is in two colors, as may be seen from our reproduction of it (Plate VII). We present also a reproduction of the first page of the treatise, with its tastefully designed headpiece and initial letter (Plate VIII). The volume is a creditable example of seventeenth century Italian printing.

The text fills 76 pages, and ends with a quaint tailpiece, which we have reproduced at the end of the translation (p. 270). The authorizations of publication fill the greater part of the next two pages. The figures follow, brought together in a single large folding plate. In this volume for the convenience of readers we have distributed the figures among three full-page plates (numbered IX,

---

[1] Translation:

    The Prodromus of Nicolaus Steno's
    Dissertation
    Concerning a solid naturally contained
    Within a solid

    To
    The Most Serene
    Ferdinand II
    Grand Duke of Tuscany

    Florence

    From the press under the sign of the star, MDCLXIX
    By order of the superiors.

[Plate VII.]

NICOLAI STENONIS
DE SOLIDO
INTRA SOLIDVM NATVRALITER CONTENTO
DISSERTATIONIS PRODROMVS.
A D

S E R E N I S S I M V M
FERDINANDVM II.
MAGNVM ETRVRIÆ DVCEM.

FLORENTIÆ
Ex Typographia sub signo STELLÆ MDCLXIX.
SVPERIORVM PERMISSV.

X, XI), reproducing them, however, in the original size,[1] and preserving the numerical order. The 'explanation of the figures' in the original edition is printed on an accompanying double-page folding inset. Page 79, not numbered, contains a list of corrections, which is far from complete.

There are two copies of the original edition in the British Museum. The copy which I examined (press mark 537. b. 1) has a plate preceding the text and a duplicate of it following the *explicatio figurarum*. A number of marginal corrections of the text have been made by some scholarly reader, for they are essential in every case. There is also a copy in the Library of the Royal Society, London; it is bound up with various tracts, and contains marginal corrections, but by a less careful hand. In this copy the plate and 'explanation of figures' are inserted in the front of the treatise.

The only copy which I have found in the United States is in the New York Public Library, and carries the stamp of the Astor Library on its title-page. It is in perfect condition and without marginal corrections.

II. REPRINTS OF THE ORIGINAL EDITION

1. *Nicolai Stenonis De Solido Intra Solidum Naturaliter Contento — Dissertationis Prodromus. Ad Serenissimum Ferdinandum II Magnum Etruriae Ducem.* Lugd. Batav. Apud Jacobum Moukee, 1679.

The copy of the Leyden reprint in the British Museum is a neatly printed duodecimo of 115 pages, followed by the *explicatio figurarum* and plate. The lower half of the plate has been torn off, and the upper part is bound upside down. Several other treatises are bound up with the *Prodromus*.

2. *Viri Celeberrimi Nicolai Stenonis Dani De Solido Intra Solidum Naturaliter Contento — Dissertationis Prodromus. Ad Serenissimum Ferdinandum II Magnum Etruriae Ducem Editio Secunda Etrusca.* Pistorii A. S. MDCCLXIII. Ex Typographio Publici. Praesidibus Permittentibus. Prostant etiam Florentiae apud Vicentium Landi Bibliopilam prope Monasterium Monachorum Cassinensium.

According to Maar,[2] the Bibliotheca Nazionale Braidense, Milan,

---

[1] Figure 17 alone is slightly reduced.   [2] *Opera Philosophica*, Vol. II, p. 355.

and the British Museum possess copies of this Pistoia reprint of 1763. The British Museum copy, the only one I have seen, is a quarto volume of 73 pages followed by a reduced reproduction of the plate. It is more compactly printed than the original edition, and page 68 has a brief index of contents. The original attestations on page 71 are followed by the *re-imprimatur* as follows: *Dominicus Bracciolini Vicarius Generalis Si Stampi Francesco Alfonso Tallinucci per S. M. C. Giudise ordinario di Pistoja.* The title-page bears the following quotation from Bacon:

Qui partes scribendi Historiam Naturalem sibi sumpserint hoc cogitent se non lectoris delectationi debere inservire; sed comparare rerum copiam et varietatem, quae veris axiomatibus conficiendis sufficiat. Par. ad Hist. Nat. et Exper. Aph. II.[1]

3. Facsimile Edition. Ed. W. Junk. No. 5: *N. Steno De Solido Intra Solidum Naturaliter Contento — Dissertationis Prodomus. Ad Serenissimum Ferdinandum II Magnum Etruriae Ducem. Florentiae,* 1669. *Exempl. No.* ——. W. Junk, Berlin N.W., Rathenower Str. 22, 1904.

The Berlin Facsimile is an exact reproduction of the original edition, by the heliotype process.

4. *De Solido Intra Solidum Naturaliter Contento — Dissertationis Prodromus. Ad Serenissimum Ferdinandum II Magnum Etruriae Ducem.* (A corrected text of the original edition.) Vilhelm Maar, *Nicolai Stenonis Opera Philosophica* (Copenhagen, Vilhelm Tryde, MCMX), Vol. II, pp. 181–227.

III. INCOMPLETE EDITION

E Dissertatione Nicolai Stenonis De Solido Intra Solidum Naturaliter Contento Excerpta In Quibus Doctrinas Geologicas Quae Hodie Sunt In Honore Facile Est Reperire. Curante Leopoldo Pilla, Florentiae, Ex Typographia Galilaeiana, 1842.

Pilla's edition of 1842 may be found in the Library of the University of Bologna, the British Museum, the Library of the Geological Society in London, and undoubtedly in other libraries. It

---

[1] Translation: 'Let those who undertake the writing of natural history reflect that they ought not to be subservient to the pleasure of the reader ... but that they ought to collect and prepare a store and diversity of things which may be sufficient for forming genuine axioms.'

SERENISSIME
MAGNE DVX.

GNOTAS regiones adeunti-
bus frequenter euenit, dum
per loca continuis montibus
aspera festinant ad vrbem
in vertice eorum sitam, vt
simul visam, simul proxi-
mam sibi arbitrentur, licet
multiplices viarum ambages
ad tedium vsque spem illorum morentur. Sola
enim proxima cacumina prospiciunt, quæ verò
eorumdem cacuminum obiectu occultantur, siue
edita collium, siue profunda vallium, siue cam-
porum plana, coniecturas eorum vt plurimum su-
perant, cum sibimet ipsis abblandiendo, locorum
interualla ex desiderio metiantur. Nec aliter se
res habet cum illis, qui ad veram rerum cogni
tionem

contains 28 pages and a reduced plate. The text is accompanied by brief notes in Latin. This edition, like that of Beaumont, is very incomplete.

IV. TRANSLATIONS

1. *The Prodromus to a Dissertation Concerning Solids Naturally Contained within Solids. Laying a Foundation for the Rendering a Rational Accompt both of the Frame and the several Changes of the Masse of the EARTH, as also of the various Productions in the same. By Nicolaus Steno.* English'd by H. O. London. Printed by J. Winter, and are to be Sold by Moses Pitt at the White-Hart in Little Brittain, 1671.

This translation is a bibliographical rarity, as are all editions of Steno. H. O. is undoubtedly Henry Oldenburg, who was elected Secretary of the Royal Society, April 22, 1663, and continued to act in that capacity until November 30, 1677, at a salary of forty pounds a year.[1]

Maar[2] mentions only the copies in the Royal Library of Copenhagen and the British Museum. There is also a copy in the Library of the Geological Society, London, and another in the Harvard University Library. It is odd that the Royal Society should not have been presented with a copy by the author; possibly one may have been presented, and lost. The British Museum copy is imperfect, the plate being lacking. The page facing the preface bears the name *Jos. Banks*. This can be no other than Sir Joseph Banks, the celebrated English naturalist, who became President of the

---

[1] See *Record of Royal Society*, 3d ed., 1912, p. 207; and Birch, *History of the Royal Society*, Vol. II (1756), p. 376. Oldenburg was succeeded by Robert Hooke. Under the date of June 25, 1667, Samuel Pepys remarks: "I was told, yesterday, that Mr. Oldenburg, our Secretary at Gresham College, is put into the Tower, for writing news to a virtuoso in France, with whom he constantly corresponds in philosophical matters; which makes it very unsafe at this time to write, or almost do any thing." And again, under date of April 30, 1669: "This morning I did visit Mr. Oldenburgh, and did see the instrument for perspective made by Dr. Wren, of which I have one making by Browne; and the sight of this do please me mightily."

For the records of Oldenburg's arrest on June 20, 1667, and his release on August 26 of the same year by orders of Charles II, see C. R. Weld, *History of the Royal Society*, London, 1848, vol. I, pp. 201–204. An account of Oldenburg's life may be found *ibid.*, pp. 259–261.

A brief notice of the H. O. edition was contributed, probably by Oldenburg himself, to the Royal Society. See *Philosophical Transactions*, vol. VI (1671), p. 2179 ff.; *Abridged Edition of Philosophical Transactions*, vol. I (1665–1672), London, 1809, pp. 605, 606.

[2] *Opera Philosophica*, Vol. II, p. 356.

Royal Society in 1778 and who is known to have bequeathed his collection of books and botanical specimens to the British Museum. The copy in the Library of the Geological Society is inscribed with the name *Rob. Dav* on the upper right-hand corner of the fly-leaf, where the last part of the name, Davis, has been worn away.

The Harvard copy is thus described by its donor, Professor J. B. Woodworth, in *Science*, Vol. 25 (1907), pp. 738, 739: "There are sixteen pages of preface with the title-page, and 112 pages of text and one plate; the size of the printed part of the page measures 2.75 inches wide by 5.5 inches high. . . . The copy in the writer's possession is bound up as a separately paged tract at the end of a small volume of the celebrated Robert Boyle's 'Essays of Effluvium, etc.,' containing also his 'Essay about the Origine and Virtue of Gems' of 1672.[1] A general title-page gives reference to Steno's work. This title-page is dated 1673.[2] All of the contained tracts appear to have been separately printed at different dates between 1671 and 1673, at which last date they were brought out in the form above described."[3]

---

[1] In this connection the following item from the *History of the Royal Society* (Vol. III, p. 55) is of interest: "Mr. O. presented from Mr. Boyle his *Essay about the Origin and Virtues of Gems*, printed at London, 1672, in 8vo."

[2] Woodworth's title-page, as reproduced in *Science*, vol. 25, p. 738, ascribes the publication of the treatise to *F. Winter*. The letter is not *F* but a quaintly formed *J*, as is clear from the reappearance of the same letter in the spelling of the word *juyces* in the "Interpreter to the Reader." Maar, *op. cit.*, Vol. II, p. 336 also printed *F*, but in a letter in reply to my contention writes: "Of course you are quite right. It is a *J* and not an *F* on the title-page of Steno's treatise."

[3] Dr. Maar is the possessor of a similar copy which is not described in his *Opera Philosophica*. The title-page reads:

ESSAYS
Of the
Strange Subtility
Determinate Nature } of EFFLUVIUMS
Great Efficacy
To which are annext
New Experiments to make FIRE
and FLAME Ponderable.
Together with
A Discovery of the Perviousness of *Glass*.
ALSO
An ESSAY, about the Origine
and Virtue of GEMS.
By the Honorable ROBERT BOYLE,
Fellow of the Royal Society.
To which is added
The PRODROMUS to a Dissertation
concerning *Solids* naturally contained with-

The translation is preceded by an address or preface bearing the title: "The Interpreter to the Reader." Since this is of unusual interest because of the writer's testimony regarding the results of the independent investigation of the nature of gems by Robert Boyle, we print it here in full:

"Reader,

"This Ingenious Piece, lately publish't in Italy, (where 'twas Printed in Latin), and thence come to the hands of the Interpreter, was thought fit to be English'd, chiefly upon this occasion, That the Stationer, that hath Printed it, did, upon Information given Him of the Valuable Contents thereof, earnestly sollicite, that it might forth-with be put into this Language; he not only conceiving, that there being now very little or no commerce between the English Book-sellers, and those of Italy, the conveyance of this Book, (as it doth of others there Printed) into England would prove very tardy; but also considering, that though within a reasonable time some Copies of it should come over, yet there would not be enough of them, to serve all sorts of curious English-Men, nor even that number of English Readers versed in the Latin Tongue, which this Considerable Discourse is like to meet with, forasmuch as it giveth very fair hopes, That by a due weighing of the particulars, therein laid down, the sagacious Inquirers into Nature may be much assisted to penetrate into the true knowledge of one of the Great Masses of the World, the EARTH, and therein to find out not only the Constitution of the Whole, but also the several Changes, and the various Productions made in the Parts thereof; as the Excellent Robert Boyle [1] hath of late Years, with great Acuteness as well as

<p style="text-align:center">in Solids Giving an Account of the Earth
and its Productions.
By <i>Nicholas Steno</i>. Englished by H. O.
London, Printed by W. G. for M. Pitt, at
the <i>Angel</i> near the little North Door
of S^t <i>Paul's</i> Church. 1673.</p>

A copy of this edition, consigned to the translator, was lost in the sinking of the "Hesperian," September 6, 1915. Another copy of the H. O. translation, now in the translator's possession, is an independent volume in its original calf binding. To the bibliographical data presented above, it may be interesting to add that the first page of the translation is adorned with the conventional emblems of the rose, fleur-de-lis, and thistle, each surmounted by the crown of Charles II as Sovereign of the United Kingdom. The book is very rare.

[1] The reference is to *Memoirs for a General History of the Air* (which, however, was not published until 1692), by Robert Boyle (1627–1691), in the edition of P. Shaw, Vol. III, Lon-

unwearied Industry, led us on a great way in the knowledge of another of the great Masses, the AIR, though the same also hath not been unmindful of considering this very subject, here treated of; forasmuch as He, before he would see or hear any thing of this Prodromus, did upon occasion candidly declare to the Author of this Version, (who bona fide here publickly attests it,)

"First, That he doth, upon several inducements, suppose, the generality of Transparent Gems or Precious Stones to have been once Liquid substances, and many of them, whilst they were either fluid, or at least soft, to have been imbraced with Mineral Tinctures, that con-coagulated with them; whence he conceiveth, that divers of the real qualities and vertues of Gems (for he doubts, most ascribed to them are fabulous) may be probably derived. And as for Opacous Gems and other Medical Stones, as Bloud-Stones, Jaspers, Magnets, Emery, etc. He esteems them to have, for the most part, been Earth (perhaps in some Cases very much diluted and soft,) impregnated with the more copious proportion of fine Metallin or other Mineral Juyces or Particles; all which were afterwards reduced into the forme of Stone by the supervenience (or the exalted action) of some already in-existent petrescent liquor or petrifick Spirit, which he supposeth may sometimes ascend in the forme of Steams; from whence may be probably deduced not only divers of the Medical Vertues of such stones, but some of their other Qualities, as Colour, Weight, etc. and also explained, How it may happen what He hath (and, he doubts not, others may have also) observed of Stones of another kind, or Marcasites, or even Vegetable and perhaps Animal substances, that have been found inclosed in solid Stones; For, these Substances may easily be conceived to have been lodged in the Earth, whilst it was but Mineral Earth or Mud; and afterwards to have been, as 'twere cased up by the supervenient Petrifick Agents that pervaded it.[1]

"Nor are these Petrescent liquors the only ones, to which he supposeth that many Fossils may owe their Origin, since he thinks,

---

don, 1725, pp. 15–98. The edition of Boyle's works by Thomas Birch (first edition, 1744, in five volumes, the second, 1772, in six) has not been accessible to me.

[1] "Of these Pretious Stones this Noble Philosopher was pleased to leave with the Publisher a Manuscript of his composure, now ready to be Printed, which he assur'd him it had been several Years ago."

H. O. refers to Boyle's *The Origin and Virtues of Gems*, published in 1672. Cf. Shaw, Vol. III, pp. 99–143, and above, p. 198.

there may be, (if one may so speak) both Metallescent and Mineralescent Juyces in the bowels of the Earth, and that sometimes they may there exist and operate under the forme of Spirits or Steams.[1]

"Beside this, we cannot but take notice here of what was intimated a good while ago in Numb. 32. of the Phil. Transactions, p. 628, viz. That Mr. Robert Hook had at that time ready some Discourses upon this very Argument, which, by reason of the many avocations he hath met with in the rebuilding of the City of London, and his attendance on the R. Society, he hath not yet been able quite to finish for the Press.

"Now this being so, that several judicious Persons do employ themselves in the inquiry after the Observables in the greater Parts of the World, there is no question but many remarkable things will be detected therein; and, (to speak more generally on this occasion,) since 'tis apparent, that the Ingenious and Diligent almost everywhere are entring more and more into Philosophical Leagues, for the discovery of the works of God and the Operations of Nature, we cannot but entertain pregnant hopes, that notwithstanding all the oppositions of Lazy and Envious Men, a good harvest of considerable and useful knowledge will be reaped in time, and thence good store of fruitful seed be ministred for large successive crops of the same kind, for the magnifying of our great Creatour, and the Enobling and benefiting of Man-kind."

2. *Extrait De La Dissertation de Nicolas Sténon sur les corps solides qui se trouvent contenus naturellement dans d'autres corps solides.* In *Collection Académique de Dijon, Partie Etrangère*, IV, 1757, pp. 377-414.

3. *Prodromus d'une dissertation sur le solide contenu naturellement dans un autre solide; extrait et traduit par M. Elie de Beaumont.* Paris, 1832.

This translation appeared in *Annales des Sciences Naturelles*, Paris, 1832, Vol. 25, pp. 337-377. The article is entitled *Fragmens géologiques tirés de Stenon, de Kazwini, de Strabon et du Boun-Dehesch.*

Copies of Beaumont's translation may be found in the libraries

---

[1] "About which he also was willing not only to shew to the Publisher several Observations and Collections of his in the forme of Discourses, but also to put them into his hands to peruse the same."

of the Royal Society and the Geological Society, London. The title-page of the edition in the Geological Society shows that Beaumont based his work on the Leyden edition of 1679. It is, however, a summary rather than a translation, emphasizing chiefly the section dealing with the origin of mountains.

4. *Nicolaus Steno Foreløbig Meddelelse Til en Afhandling Om Faste Legemer, Der Findes Naturlig Indlejrede I Andre Faste Legemer I Oversættelse Ved August Krogh Og Vilhelm Maar Med Indledning Og Noter*, København, MCMII. Gyldendalske Boghandels Forlag Langkjærs Bogtrykkeri.

This is a quarto volume containing a portrait of Steno, an Introduction, pp. i–xii, the Translation, pp. 3–89, the Explanation of Figures, pp. 91–93, and a half-sized reproduction of the original plate. The attestations are given on page 97, not numbered, and page 98. Pages 101–106 contain notes. The edition is limited to 700 numbered copies.

V. SELECTED REFERENCES

Angelis, de, Article *Sténon* in *Biographie Universelle (Michaud), Ancienne et Moderne*, Nouvelle Edition, Tome Quarantième, Paris, pp. 209–211.

Capellini, G., *Di Nicola Stenone e dei suoi studi geologici in Italia*, University of Bologna, 1870.

Chéreau, Article *Sténon* in *Dictionnaire Encyclopédique des Sciences Médicales*, Troisième Série, Tome Onzième, Paris, 1883, pp. 689–691.

Eloy, N., *Dictionnaire Historique de la Médicine*, Tome Second, Liège, 1755, pp. 391–393.

Fabronius, A., *Vitæ Italorum Doctrina Excellentium Qui Saeculis XVII et XVIII Floruerunt* (Pisis, 1778–1805), Vol. III (1779).

Geikie, A., *The Founders of Geology* (2d ed., London, 1905), pp. 53–60.

Gosch, C. C. A., *Udsigt over den danske Zoologiske Literatur*, 2 Afdeling, 1 Hefte, København, 1872.

Hughes, T. M., *Steno*, in *Nature*, Vol. 25 (1882), pp. 484–486.

Huxley, T. H., *The Rise and Progress of Palæontology, Discourse at York* (Meeting of the British Association), in *Nature*, Vol. 24 (1881), pp. 452-455.

Jorgensen, A. D., *Niels Stensen*, København, 1884.

Köcher, A., *Herzog Johann Friedrich, Bischof Steno, u. Pastor Petersen*, in *Zeitschrift des historischen Vereins für Niedersachsen*, 1889, pp. 204-212.

Lorenzen, A., *Niels Stensen, Der Vater der Geologie*, in *Die Natur*, Vol. 3 (1854), p. 220 ff.

Lyell, C., *Principles of Geology* (9th ed., New York, 1853), pp. 21-24.

Manni, D. M., *Vita del Litteratissimo Mgr. Niccolo Stenone*, Firenze, 1775.

Plenkers, W., *Der Däne Niels Stensen, Ein Lebensbild nach den Zeugnissen der Mit- und Nachwelt entworfen*, Freiburg im Br., 1884.

Spencer, L. J., Article *Crystallography*, in *Encyclopædia Britannica*, 11th ed., Vol. VII, pp. 569, 570; cf. also Article *Steno*, *ibid.*, Vol. XXV, p. 879.

Sollas, W. J., *The Influence of Oxford on the History of Geology*, in *Science Progress*, Vol. 7 (1898), pp. 25-29.

Wichfeld, J., *Erindringer om den Danske Videnskabsmand Niels Stensen*, in *Historisk Tidsskrift*, 3 Række, 4 Bind (Kjøbenhavn, 1865), pp. 1-109.

Woodworth, J. B., *Steno*, in *Science*, Vol. 25 (1907), pp. 738, 739.

Von Zittel, K. A., *Geschichte der Geologie und Paläontologie bis Ende des 19 Jahrhunderts* (München u. Leipzig, 1899), pp. 32-36.

Steno ist der erste Forscher welcher geologische Probleme auf inductivem Wege zu lösen versuchte und zugleich eine klare Vorstellung davon hatte, dass die Geschichte der Erde aus ihrer Zusammensetzung und ihrem Aufbau ermittelt werden könne. Für die Entwickelung der Geologie blieben leider die Schriften dieses Scharfsinnigen Forschers ohne jegliche Bedeutung; sie wurden von den Zeitgenossen kaum beachtet, geriethen in Vergessenheit und fanden erst in diesem Jahrhundert durch Elie de Beaumont und Alexander von Humboldt die verdiente Anerkennung.

— *Von Zittel, Geschichte der Geologie und Paläontologie*, pp. 35, 36.

THE PRODROMUS

Most Serene Grand Duke:[1]

Travellers into unknown realms frequently find, as they hasten on over rough mountain paths toward a summit city, that it seems very near to them when first they descry it, whereas manifold turnings may wear even their hope to weariness. For they behold only the nearest peaks, while the things which are hidden from them by the interposition of those same peaks, whether heights of hills, or depths of valleys, or levels of plains, far and away surpass their guesses; since by flattering themselves they measure the intervening distances by their desire.

P. 2.[2] So, and not otherwise, is it with those who proceed to true knowledge by way of experiments; for as soon as certain tokens of the unknown truth have become clear to them, they are of a mind that the entire matter shall be straightway disclosed. And they will never be able to form in advance a due estimate of the time which is necessary for loosing that knotted chain of difficulties which, by coming forth one by one, and from concealment, as it were, delay, by the constant interposition of obstacles, them that are hastening toward the end. The beginning of the task merely reveals certain common, and commonly known, difficulties, whereas the matters which are comprised in these difficulties—now untruths which must be overthrown, now truths which must be established; sometimes dark places which must be illumined, and again, unknown facts which must be revealed—shall rarely be disclosed by any one before the clew of his search shall lead him thither. Democritus,[3] not un-

---

[1] Ferdinand II, Grand Duke of Tuscany; see p. 179.

[2] The pagination is that of the original publication, which is reproduced in the Berlin Facsimile (p. 196).

[3] Steno doubtless had in mind the proverb recorded by Diogenes Laertius (IX. 72); ἐτεῇ δὲ οὐδὲν ἴδμεν· ἐν βυθῷ γὰρ ἡ ἀλήθεια, 'In reality we know naught, for truth lies in a well.' βυθός, strictly speaking, denotes the depth of the sea (cf. Æschylus, *Prometheus*, 432). It is in this sense that Cicero (*Academica Prior.*, ii. 10, 32) repeats the proverb: *naturam accusa, quae in profundo veritatem, ut ait Democritus, penitus abstruserit,* 'Accuse nature, which has

wisely, was wont to use the illustration of a well, wherein one could scarcely estimate aright the task and time of draining it dry, except by draining it dry, since both the number and the volume of the hidden springs leave the amount of the intake a matter of doubt.

Do not be surprised, therefore, Most Serene Prince, if, for a whole year's time, and, what is more, almost daily, I have said that the investigation for which the teeth of the shark[1] had furnished an opportunity, was very near an end. For having once or twice seen regions where shells and other similar deposits of the sea are dug up, when I observed that those lands were sediments of the turbid sea and that an estimate could be formed of how often the sea had been turbid in each place, I not only over-hastily fancied, but also dauntlessly informed others, that a complete investigation on the spot was the work of a very short time. But thereafter, while I was examining more carefully the details of both places and bodies, these day by day presented points of doubt to me as they followed one another in indissoluble connection, so that I saw myself again and again brought back to the starting-place, as it were, when I thought I was nearest the goal. I might compare those doubts to the heads of the Lernean Hydra, since when one of them had been got rid of, numberless others were born; at any rate, I saw that I was wandering about in a sort of labyrinth, where the nearer one approaches the exit, the wider circuits does one tread.[2]

But I shall not tarry to excuse this tardiness of mine, since it is abundantly evident to you, from long experience, how per-

---

hidden the truth completely, as Democritus says, in the depth.' Cf. also *Acad. Post.*, i. 12, 44: *Democritus (dixit) in profundo veritatem esse demersam.*

Steno's use of the word *puteus* ('well') accords with the expression of the proverb in his time. Rabelais (*Pantagruel*, iii. 36) incorrectly ascribes the saying to Heraclitus: *Je suis descendu au puiz tenebreux, auquel disoit Heraclitus estre vérité cachée.* A curious addition to the original was made by Francis Bacon (edition of Spedding, Ellis, Heath, Vol. XIII, p. 383) when he wrote, "Democritus said 'that truth did lie in profound pits, and when it was got, it needed much refining.'"

The proverb has given rise to several allegorical paintings of Truth in which the well figures prominently. Among these are Paul Baudry's "Truth," J. J. Lefebure's "Truth," both in the Luxembourg; and Titian's so-called "Sacred and Profane Love," in the Borghese Gallery.

[1] *Canis Carcharia.* Steno's treatise *Canis Carchariae Dissectum Caput* is dated 1667 and is reprinted by Maar, *N. Stenonis Opera Philosophica*, Vol. II, pp. 113-145. Cf. p. 125, especially.

[2] The language is reminiscent of Seneca, *Epistles*, 44. 7.

plexing is a matter which is involved in a series of experiments. But the fact that, after a large part of the task assayed had been completed, I should drop everything and ask your leave to return to my native land to pursue anatomical investigation — this indeed would demand an excuse did I not know that this obedience on the part of a subject of another prince would not be displeasing to you, which in a similar circumstance would please you on the part of your own subjects. And this hope of mine concerning your kindness is made surer by that exceptional goodwill, whereby, through devoting generous assistance[1] to the advancement of my studies, you wished that unrestricted opportunity for learning should be left to me whenever occasion might arise. Therefore, since I no longer dare to hope for the time necessary for finishing the tasks which I have begun, I shall do in the payment of my promises what has been conceded by common custom to debtors; when they have not the means to pay in full, they pay what they have, in order that they may not be forced to withdraw from business. Since, then, I am unable to complete all the things which were to be shown to you, I shall offer the chief of what I have found, in order that I may not appear to have deceived you.

P. 4.

I should gladly have postponed everything until it had been possible for me, on my return to my native land, to perfect the details, were I not awaiting the same fortune there which I have hitherto experienced everywhere, in that new tasks have constantly stood in the way of finishing those first undertaken. While I was intent upon counting the glands of the entire body,[2] the wonderful structure of the heart[3] carried me away into an examination of it; and the deaths of my kin[4] interrupted the studies I had begun on the heart. Your seas furnished us a shark[5] of marvellous size to keep me from applying myself to

---

[1] See p. 179.   [2] For Steno's work on the glands, see pp. 176, 188 f.

[3] Steno first mentions his study of the heart in a letter to Thomas Bartholin dated "the last of April," 1663, Leyden. See Maar, *Opera Philosophica*, Vol. I, p. 155. In 1667 Steno published his *Elementorum Myologiae Specimen, Seu Musculi Descriptio Geometrica.* Cf. p. 190.

[4] While studying in Leyden, 1664, Steno learned of the death of his step-father, Johannes Stichman. The death of Steno's mother occurred soon after his arrival in Copenhagen. See p. 178.

[5] For the treatise *Canis Carchariae Caput*, see p. 206, note 1. Compare also *Historia Dissecti Piscis Ex Canum Genere*, Maar, *op. cit.*, Vol. II, pp. 147–155.

the detailed description of the muscles; and now, when I am wholly devoted to my present experiments, he whose command the law of nature bids me heed, and whose great kindness toward me and mine constrains me, calls me to other things.[1]

To what end all these matters may come, I do not care to inquire anxiously, lest it be, peradventure, to accredit to myself things which are due to a higher cause. If long contemplation had added something of my own, as it were, to discoveries not my own, certainly if I had tarried longer in working out one discovery, I should myself have shut the door to the finding of the rest. And so, not knowing what other experiments and studies may await me elsewhere, I thought it best to set forth here these matters *concerning a solid naturally contained within a solid*, which shall be a pledge to you of gratitude for the favors I have received and shall afford to others, who are enjoying their desired leisure, an opportunity of pursuing their studies of physics and geography with greater profit.

As regards the production of a solid naturally contained within a solid, I shall first sketch briefly the method of my Dissertation, then explain concisely the more noteworthy matters which appear in it.

The Dissertation itself I had divided into four parts, of which the first, taking the place of an introduction, shows that the inquiry concerning sea objects found at a distance from the sea, is old, delightful, and useful; but that its true solution, less doubtful in the earliest times, in the ages immediately following was rendered exceedingly uncertain. Then after setting forth the reasons why later thinkers abandoned the belief of the ancients, and why, although one may read a great many excellent works by many authors, the question at issue has hitherto been settled by no one anew,[2] I show, returning at

---

[1] Steno refers to the invitation of Frederik III; cf. p. 181. He got no farther than Amsterdam, however. See p. 182.

[2] There is an interesting discussion on the nature and origin of fossil shells in Bernard Palissy's *Des Pierres* (*Discours Admirables*, etc., 8vo, Paris, 1580). Referring to Palissy, Fontenelle (*Histoire de l'Académie des Sciences*, Année 1720, p. 5) remarks: "Un potier de terre, qui ne savait ni latin ni grec, fut le premier qui, vers la fin du XVIe siècle, osa dire dans Paris, et à la face de tous les docteurs, que les coquilles fossiles étaient de véritable coquilles déposées autrefois par la mer dans les lieux où elles se trouvaient alors, que des animaux, et surtout des poissons, avaient donné aux pierres figurées toutes leurs différentes figures; et il

length to you, that besides very many other things which under your auspices have in part been discovered, and in part freed from old doubts, to you is due our trust that the finishing touch shall soon be put upon this investigation also.

In the second part is solved a universal problem upon which depends the unravelling of every difficulty, and it is this: *given a substance possessed of a certain figure, and produced according* P. 6. *to the laws of nature, to find in the substance itself evidences disclosing the place and manner of its production.* In this connection, before I proceed to unfold the solution of the problem, I shall strive to expound all its terms, with the view of leaving no school of philosophers in doubt, and in dispute, as to their significance.

The third part I have reserved for the investigation of different solids contained within a solid, in accordance with the laws discovered in the solution of the problem.

The fourth part describes various conditions in Tuscany not treated by historians and writers upon natural subjects, and sets forth a process of the universal deluge which is not at variance with the laws governing movements of nature.

I had indeed begun to set forth these things in Italian, both because I knew this would please you, and in order that it might appear to the illustrious Academy[1] which has enrolled

---

défia hardiment toute l'école d'Aristote d'attaquer ses preuves." The quotation is taken from Flourens, *De la Longévité humaine et de la Quantité de Vie sur le Globe*, Paris, 1855, pp. 200, 201.

The dialogue in Palissy's *Discours* is between *Theorique* and *Practique*, whose contention may be illustrated by the following quotation:

"Et par ce qu'il se trouue aussi des pierres remplies de coquilles, iusques au sommet des plus hautes montagnes, il ne faut que tu penses que lesdites coquilles soyent formees, comme aucuns disent que nature se ioue à faire quelque chose de nouueau. Quand i'ay eu de bien pres regardé aux formes des pierres, i'ay trouué que nulle d'icelles ne peut prendre forme de coquille ny d'autre animal, si l'animal mesme n'a basti sa forme: parquoy te faut croire qu'il y a eu iusques au plus haut des montaignes des poissons armez et autres, qui se sont engendrez dedans certains cassars ou receptacles d'eau, laquelle eau meslee de terre e d'un sel congelatif et generatif, le tout s'est reduit en pierre auec l'armure du poisson, laquelle est demeuree en sa forme. . . . Il faut donc conclure que auparauant que cesdites coquilles fussent petrifiées, les poissons qui les ont formées estoyent viuans dedans l'eau qui reposoit dans les receptacles desdites montagnes, et que depuis l'eau et les poissons se sont petrifiez en un mesme temps, et de ce ne faut douter." *Œuvres Complètes de Bernard Palissy*, by Paul-Antoine Cap, Paris, 1844, pp. 277, 279.

[1] The Accademia del Cimento; see p. 180. This Academy came to an end in 1667 when its founder, Leopold de' Medici, became a Cardinal.

me among its members, that as I am least worthy of so great an honor, so am I most desirous of proving the attempts whereby I am striving to attain some knowledge of the Tuscan tongue. But I am not grieved that the necessity has been placed upon me of postponing that writing; for as my present journey promises me a fuller knowledge of matters serving to elucidate my investigation, so the delay assures me of a happier advancement in my study of the language.

It would be a long task to write out in detail all my observations, together with the conclusions drawn from them, developed in accordance with the method suggested; wherefore, I shall report sometimes conclusions, and again observations, as may seem best, in order to explain the chief points briefly, and as clearly as possible.

P. 7.

The reason why, in the solution of natural questions, not only do many doubts remain undecided but, for the most part, such doubts multiply with the number of writers, seems to me to depend chiefly upon two causes.

The first cause is that few take it for granted that all those difficulties, without whose solution the settlement of the question itself is left marred and incomplete, must be removed. The present inquiry illustrates this point. Only a single difficulty vexed the ancients, that is in what way marine bodies had been left in places far from the sea; and the question was never raised whether similar substances had been produced elsewhere than in the sea.[1]

---

[1] The presence of fossil shells in places remote from the sea is discussed in the *Geography* of Strabo (*circa* 67 B.C.–19 A.D.), C. 49, 50 (I. 3, 4):

'He (Eratosthenes) says that a particularly interesting subject of inquiry is afforded by the fact that an abundance of cockle, oyster, and scallop shells, and salt-water lakes are frequently seen far inland, two or three thousand stadia from the sea, as in the case of the temple of Ammon and the road leading up to it for a distance of three thousand stadia. For a profusion of oyster shells, salt beds, and salt springs can still be found there at the present time. In addition to this, wrecks of sea-going vessels are pointed out which were said to have been cast up through some chasm. . . .

'In this he agrees with the opinion of Strato, the physicist, and of Xanthus, the Lydian. Xanthus asserts that in the reign of Artaxerxes there was so great a drought that the rivers, lakes, and wells dried up; and that he had frequently found, far from the sea, fossil shells, some like cockles and others like scallops, as well as salt lakes in Armenia, Matiana, and Lower Phrygia. On this account he was convinced that what are now plains had once been sea. Strato, who searched more deeply for the causes of these phenomena, believed that the Euxine formerly had no outlet at Byzantium, but the rivers which emptied into the Euxine had forced an opening, and that the water thereupon fell into the Propontis and the Hellespont.

In more recent periods the difficulty of the ancients was pressed with less insistence, since almost all were concerned with tracing out the origin of the bodies mentioned. They who ascribed them to the sea accomplished this result: they proved that bodies of this character could not have been produced by any other agency. They who attributed these bodies to the land, denied that the sea could have covered the places where they were found, and were wholly engaged in praising the forces of a Nature of which they had little knowledge — forces fitted to produce anything whatsoever. Perhaps a third opinion may properly be admitted, in accordance with which a part of the bodies under consideration is regarded as attributable to the land, and a part to the sea. Nevertheless there is deep silence almost everywhere concerning the doubt of the ancients, except that some make mention of floods, and a sort of immemorial succession of years, but merely in passing and, as it were, in

P. 8. treating another subject. In order, therefore, to satisfy, to the best of my ability, the laws of analysis, I wove and unwove the web of this research, and scrutinized its various details again and again, until I saw no difficulty left any longer in the reading of authors, or in the objections of friends, or in the examination of places, which I had not either solved, or at least determined, from facts hitherto known to me, to what extent a solution was possible.

The first question was, whether *Glossopetrae Melitenses*[1] were

'The same thing happened in the case of the Mediterranean. For the sea, after having been filled by the rivers emptying into it, had broken a passage through at the Pillars (Gibraltar), and the places formerly covered with shoal water, were left dry by this eruption. He (Strato) finds the cause for this, first, in the fact that the bottoms of the Atlantic and of the Mediterranean are not on the same level, and, secondly, in the fact that even now a sand-bank runs beneath the water from Europe to Libya, bearing witness to the time when the Mediterranean and the Atlantic were not united. Strato also said that the waters around Pontus are very shallow, whereas off Crete, Sicily, and Sardinia they are very deep. . . .

'Egypt, too, he said, was formerly covered by the sea as far as the marshes near Pelusium (Tineh) and Mount Casius (El-Kas) and Lake Sirbonis (Lake Sebaket-Bardoil). Even now, when salt is dug in Egypt, the beds are found beneath layers of sand and mixed with fossil shells, as if the country had formerly been under the sea, and all the region around Casium and Gerra (Maseli) had a shoal extending to the Arabian Gulf.'

[1] The literal translation of *Glossopetrae Melitenses* is 'tongue-stones from Malta.' In the treatise *Canis Carchariae Dissectum Caput* (1667), Steno was not free from doubt as to the origin of the 'stones,' as shown by the following passage (Maar, *op. cit.*, Vol. II, pp. 127, 128):

No decision has yet been reached regarding the larger *glossopetrae*, as to whether they are shark's teeth or stones formed in the earth. Some have maintained that substances found in

once the teeth of sharks; this, it was at once apparent, is identical with the general question whether bodies which are similar to marine bodies, and which are found far from the sea, were once produced in the sea. But since there are found also on land other bodies resembling those which grow in fresh waters, in the air, and in other fluids, if we grant to the earth the power of producing these bodies, we cannot deny to it the possibility of bringing forth the rest. It was necessary, therefore, to extend the investigation to all those bodies which, dug from the earth, are observed to be like those bodies which we elsewhere see growing in a fluid. But many other bodies, also, are found among the rocks, possessed of a certain form; and if one should say that they were produced by the force of the

the earth, resembling parts of animals, are the remains of animals which once lived on the spot; while others believe that such substances were formed in the earth without reference to animals. I have not sufficient knowledge in these matters to venture an opinion at this time. Although my travels [Steno accompanied Ferdinand II in his travels through Tuscany; cf. Plenkers, *Niels Stensen*, pp. 31, 58] have led me through various regions of this sort, I would not presume to assert that the places which I shall see in the rest of my journey will correspond to those I have already examined; especially since I have not yet seen the regions which my distinguished teacher Bartholin has examined in his journey in Malta. Just as in court, therefore, one man takes the rôle of defendant, another of plaintiff, while both submit to the decision of the judge, so I shall present, as a result of my observations, the reasons for ascribing such substances to animals. At another time I may set forth the reasons for the opposite belief, but I shall always await a true decision from those who are better informed.' For Steno's illustrations of *glossopetrae* see Maar, *op. cit.*, Vol. II, Tab. III.

Pliny (*Natural History*, XXXVII. 164) states that the *glossopetra*, resembling the human tongue, 'is not produced (*nasci*) in the earth, as tradition relates, but falls from heaven at the time of the waning moon.' Compare, further, O. Abel, article *Paläontologie und Paläozoologie*, in *Die Kultur der Gegenwart, Dritter Teil, Vierte Abteilung, Vierte Band* (Leipzig, 1914), pp. 313, 314:

"Albertus Magnus hatte noch die Möglichkeit zugegeben, dass die Versteinerungen nicht ausschliesslich Produkte der Virtus formativa seien, sondern dass auch die Leichenreste fossiler Tiere und Pflanzen dort zu Stein werden könnten, wo eine steinmachende Kraft ihren Einfluss ausüben könne. Ungefähr in denselben Bahnen bewegen sich die Vorstellungen von Georg Bauer, genannt Agricola (1494-1555); Haifischzähne, die er nach dem Vorbilde des älteren Plinius 'Glossopetren' nennt (eine Bezeichnung, die noch G. W. Leibniz für fossile Pottwalzähne 1749 gebrauchte), sind nach Agricola 'verhärtete Wassergemenge.'"

Ibid., p. 344: "Die Glossopetren des Plinius sind fossile Haifischzähne, und da solche in tertiären Bildungen zu den häufigeren Wirbeltierresten gehören, so erregten sie schon frühzeitig die Aufmerksamkeit. Es berührt eigentümlich, noch heute einen einfachen Taglöhner, der gewiss nicht die mindeste Ahnung von der Literatur der Scholastenzeit besitzt, einen fossilen Haifischzahn, den er in seinem Bruche fand, als 'Vogelzunge' bezeichnen zu hören. Bis zur Zeit Knorrs und Walchs gingen Haifischzähne vorwiegend unter der Bezeichnung Zungensteine, Vogelzungen, Schwalbenzungen oder Schwalbensteine, Lamiodonten, Schlangenzungen usw. durch die Literatur, und noch Leibniz hielt an der Bezeichnung 'Glossopetra' des älteren Plinius fest."

place, one must confess that all the rest were produced by the same force. And so I saw the matter finally brought to the point that any given solid naturally contained within a solid must be examined in order to ascertain whether it was produced in the same place in which it is found; that is, the character

P. 9. not only of the place where it is found, but also of the place where it was produced, must be investigated. But no one, in truth, will easily determine the place of production who does not know the manner of production, and all discussion concerning the manner of production is idle unless we gain some certain knowledge concerning the nature of matter. From this it is clear how many questions must be solved in order that a single question may be set at rest.

The second cause, the nurse of doubts, seems to me to be the fact that in the consideration of questions relating to nature those points which cannot be definitely determined, are not distinguished from those which can be settled with certainty. And the result is that the leading schools of philosophers are reduced to two classes. Some religiously refrain from putting credence even in demonstrations, out of fear that there be lurking in them the error which they have often found in other asserted truths. Others, on the contrary, would by no means allow themselves to be bound to consider as certain only those matters to which no one of sound mind and sound senses could deny credence, but believe that all things are true which have seemed to themselves fine and clever. Nay, the very advocates of experiment have rarely had sufficient self-control to refrain either from casting aside even most certain fundamental facts of nature, or from considering the fundamental facts discovered by themselves as proved. In order that I might, therefore, avoid[1] this rock also, I decided that in the sciences of nature we must enforce the principle which Seneca[2]

---

[1] In the Florentine edition of 1669, *evitare* is an obvious misprint for *evitarem*.

[2] Seneca nowhere, so far as I know, gives expression to the first part of Steno's sentence. But the language of the last part is identical with that of *Epistles*, 29. 11 : *ex omni domo conclamabunt, Peripatetici, Academici, Stoici, Cynici*. The only point of difference is that Seneca has the future tense, *conclamabunt*, where Steno has the present, *conclamant*. The first part of Seneca's sentence does indeed mention 'the people,' but scarcely in a manner apropos of Steno's argument. Epicurus is represented as saying: 'I never wished to please the people. For the people does not approve what I know, and what the people approves, I do not know.' Steno may have been thinking of *de Beneficiis*, I. 11, 1, where benefits are classified as neces-

again and again inculcates concerning maxims of morality; he says that those are the best maxims of morality which are common, which are of the people, which all of every school proclaim, Peripatetics, Academics, Stoics, and Cynics. And certainly those statements of the fundamental facts of nature cannot fail to be best which are common, which are accepted of the people, which all of every school are held to acknowledge, both those who in everything are desirous of novelty, and those who are devoted to ancient doctrines.

I do not determine, therefore, whether the particles of a natural body, can or cannot undergo change, as its form does; whether there are or are not minute interstices; whether in those particles there is present, besides extension and hardness, something else unknown to us; for these expressions are not of common acceptance, and it is a weak argument to deny that there is anything else in a certain object because I do not discern anything else in it.

I do assert, however, without hesitation:

1. That a natural body is an aggregate of imperceptible particles which is subject to the operation of forces proceeding from the magnet, fire, and sometimes light also; in whatever way, indeed, passages may be found, whether between the particles, or in the particles themselves, or in both.

2. That a solid differs from a fluid in that in a fluid the imperceptible particles are in constant motion, and mutually withdraw from one another; while in a solid, although the imperceptible particles may sometimes be in motion, they hardly ever withdraw from one another so long as that solid remains a solid and intact.

3. That while a solid body is being produced, its particles are in motion from place to place.

4. That as yet we know of nothing in the nature of matter by the aid of which the principle of motion, and the perception of motion, can be explained; but that the determination of natural motions can be altered by three causes:

(1) By the motion of a fluid permeating all bodies; and we

sary, useful, and pleasant, and of *Ep.*, 12. 11, where 'the best things' are said to be 'common': *quae optima sunt, esse communia.*

say that those things which are produced in this way are produced naturally.

(2) By the motion of living beings; and many of those things which in this way are produced by man, are said to be artificial.

(3) By the first and unknown cause of motion; and even the pagans believed that there was something divine in motions which originate in this way. Surely to deny to this cause the power of accomplishing results contrary to the usual course of nature, is the same as to deny to man the power of changing the courses of rivers; or of battling with sails against the winds; or of kindling fire in places where without man fire would never be kindled; or of quenching a light which would not wane except with the ceasing of its supply; or of ingrafting the shoot of one plant upon the branch of another; or of producing summer fruits in mid-winter months; or of making ice in the very heat of summer; or of a thousand other things of the kind which are in conflict with the usual laws of Nature. For if we ourselves, who know not the structure of our own and other bodies, change the determination of natural motions every day, why cannot He change their determination who not only knows our structure and that of all things, but also made them? To be ready, again, to marvel at the cleverness of man acting with free will in the case of things done by his skill, and to deny a free agent to the products of Nature, would indeed seem to me to betoken great lack of penetration; since when man has performed the most ingenious things, he cannot, save through a cloud, discern what he has done, or what instrument he has used, or what that cause is by which the instrument is moved.

These details, proved both by experiments and by arguments, I shall set forth at greater length in the Dissertation itself, in order that it may be clear that there is no philosopher who does not say the same thing, although not always in the same words; or, if he has said otherwise, who does not, nevertheless, agree to the principles from which these details necessarily follow. For the statements I have affirmed concerning matter hold true in all cases, whether one considers matter as atoms, or particles changeable in a thousand ways, or the four elements, or whatsoever chemical elements may be assumed to suit the differences of opinion among chemists. And further, the statements which I have made concerning the determination of motion, are consistent with every agent, whether you call the agent the form, or the qualities proceeding from the form, or the idea, or the tenuous common substance, or the tenuous particular substance, or the individual soul, or the world soul, or the immediate act of God.

After these things I shall explain the various modes of speaking admitted by common usage, whereby we explain in different ways the different production of different, and sometimes the same, substances; for whatever contributes anything to the production of any substance, does this either as place, or as matter, or as the agent. Hence when like produces like, it contributes to that object the place, the matter, and the motion of production, just as the small plant enclosed within the seed of some plant receives from that parent plant the matter in which it has been produced, the matter from which it has been produced, and the motion of the particles by which it has received its form; this same thing is true of animals enclosed within the egg of similar animals.

While the particular form or soul is producing something, the motion of particles in the production of that body is determined by some particular agent, whether this be the agent of another similar body or something else similar to this agent.

The things which are said to be produced by the sun receive the motion of their particles from the sun's rays, just as those which are attributed to the influences of the stars may receive the motion of their particles from the stars; for since it is certain that our eyes are affected by the light of the heavenly bodies, it will also be beyond cavil that the rest of matter may be affected by them in the same way.

The things which the earth bears receive from the earth nothing except the place in which they are produced and the matter supplied to them through the pores of the place.

P. 14. The things which are produced by Nature receive the motion of their particles from the motion of a penetrating fluid, whether this fluid come from the sun, or from the fire contained within the matter of the earth, or from some other cause unknown to us, as the agency of the soul, and so on.

He, therefore, who attributes to Nature the production of any thing, names the universal agent which appears in the production of all things; he who calls the sun to share, limits that agent a little more; he who names the soul or the particular form, mentions a more limited cause than the rest: but one who nevertheless duly weighs the answers of all, finds nothing known, seeing that Nature, the sun's rays, the soul, and the particular form, are things known only by name. But since, besides the agent, matter and place ought to be taken account of in the production of substances, it is clear that the answer ("produced by Nature") is not only more unknown than the very thing under investigation, but altogether incomplete; as, for example, mollusks found on land are said to have been produced by Nature, while those that grow in the sea are also Nature's work. Nature indeed produces all things, seeing that the penetrating fluid has a place in the production of all things; but one may also say with truth that Nature produces nothing, since that fluid by itself accomplishes nothing; its determination depends upon the place and the matter to be moved. We find an illustration in man: he can produce anything if all the necessary things are at hand, but if they are wanting, can produce nothing.

P. 15. He who attributes the production of anything to the earth, names the place indeed, but since the earth affords place, in part at least, to all the things of earth, the place alone does not account for the production of the body. The same thing can be said about the earth as about Nature; that is, the things which are formed in the earth are all produced by the earth, and of those things which are formed in the earth none is produced by the earth.

The few points set forth above suffice for the solution of all the doubtful issues in our inquiry, and I have desired to sum them up now in the three following propositions:

I

If a solid body is enclosed on all sides by another solid body, of the two bodies that one first became hard which, in the mutual contact, expresses on its own surface the properties of the other surface. Hence it follows:

1. That in the case of those solids, whether of earth, or rock, which enclose on all sides and contain crystals, selenites, marcasites,[1] plants and their parts, bones and the shells of animals, and other bodies of this kind which are possessed of a smooth surface, these same bodies had already become hard at the time when the matter of the earth and rock containing them was still fluid. And not only did the earth and rock not produce the bodies contained in them, but they did not even exist as such when those bodies were produced in them.

P. 16.

2. That if a crystal is enclosed in part by a crystal, a selenite by a selenite, a marcasite by a marcasite, those contained bodies had already become hard when a part of the containing bodies was still fluid.

3. That in the earth and rock in which crystalline and petrified shells, veins of marble, of lapis lazuli, silver, mercury, antimony, cinnabar, copper, and other minerals of this kind are contained, the containing bodies had already become hard at the time when the matter of the contained bodies was still fluid; and that, consequently, the marcasites were produced first, then the stones in which the marcasites are enclosed, and, finally, the veins of minerals which fill the fissures of the rock.

II

If a solid substance is in every way like another solid substance, not only as regards the conditions of surface, but also as regards the inner arrangement of parts and particles, it will also be like it as regards the manner and place of production, if you except those conditions of place which are found time

[1] By the term "crystals," Steno means mineral quartz (cf. p. 237); *selenites* refers to crystals of gypsum, and *marcasites* to pyrites (cf. p. 225, n. 1).

and again in some place to furnish neither any advantage nor disadvantage to the production of the body. Whence it follows:

P. 17.   1. That the strata of the earth, as regards the place and manner of production, agree with those strata which turbid water deposits.

2. That the crystals of mountains, as regards the manner and place of production, agree with the crystals of niter,[1] although it is not therefore essential that the fluid in which they were produced should have been aqueous.

3. That those bodies which are dug from the earth and which are in every way like the parts of plants and animals, were produced in precisely the same manner and place as the parts of the plants and the animals were themselves produced. But in order that no uncertain interpretation of the term place may beget new doubts, I shall forestall that difficulty here.

By the term place I mean that matter which with its own surface is in immediate contact with the surface of the body which is said to be in that place. But that matter allows sundry differences, for:

(1) It is either wholly solid, or wholly fluid, or partly solid and partly fluid.

(2) It is either wholly perceptible by itself, or partly perceptible by itself and partly perceptible through tests.

(3) It is either wholly contiguous to the body which it contains within itself, or even partly continuous with the same body.

(4) It is either always the same or undergoes change gradually. Thus the place in which a plant is produced is the matter, like that of the plant, within which the minute plant receives its form. Thus the place in which the plant grows is all that matter which, with its own surface, is in immediate contact with the entire surface of the plant, consisting sometimes of
P. 18.   earth and air, sometimes of earth and water, sometimes of earth, water, and air; sometimes of only stone and air, as in underground places I have time

[1] Steno was, of course, ignorant of the chemical difference between quartz and niter. The first is silicon dioxide, SiO_2, and the second is saltpeter, $NaNo_3$.

and again seen the roots of small plants, without any covering of earth at all, clinging to the surface of the tuff. Thus the place where the orange grows, after the blossom has fallen, is partly the peduncle continuous with it, and partly the air contiguous to it. Thus the place where the first growth occurs in animals, is partly the amniotic fluid contiguous to it, and partly the umbilical vessels scattered through the chorion, continuous with it.

III

If a solid body has been produced according to the laws of nature, it has been produced from a fluid.

In the production of a solid body, both its first outlines and its growth should be taken into account; but as I freely confess that the outline of most of the bodies is not only doubtful to me, but wholly unknown, so do I believe, without any hesitation, that nearly all the following statements concerning their growth are true.

A body grows while new particles, secreted from an external fluid, are being added to its particles. This addition, moreover, takes place either immediately from an external fluid, or through one or more mediating internal fluids.

P. 19. The additions which are made directly to a solid from an external fluid sometimes fall to the bottom from their own weight, as sediments do; sometimes the additions are made to a solid on all sides from a fluid bearing matter toward the solid, as in the case of incrustations; or the additions are made only to certain places of the solid surface, as in the case of those bodies which present fibres, branches, and angular bodies.[1] Here it must be noted in passing that the processes mentioned sometimes continue until an entire cavity is filled with such additions, and hence replacements occur which are sometimes simple, are sometimes formed from incrustations, sometimes from sediments, sometimes from angular bodies, and sometimes from various bodies variously intermingled.

The particles which are added to a solid by a mediating

[1] *Angulata corpora* is the phrase used by Steno to denote crystals in general; *crystallus* is confined to quartz. See p. 218, n. 1.

internal fluid either take on the form of fibres (since they are partly added through open pores along the length of the extended fibril, and are partly disposed in the interstices of the fibrils into the form of a new fibril by the permeating fluid), or form simple replacements: and in these two ways plants and animals are formed. Since I am less familiar with the anatomy of plants, I do not decide whether there are present several internal fluids; but it is certain that in animals different internal fluids are to be found, and I shall try to reduce these to a definite classification.

Besides the attenuated fluid permeating all things, we note in the case of animals at least three kinds of fluids, of which the first is external; the second is internal and common, the third an internal fluid peculiar to each part. By the term external fluid I mean that fluid in animals which not only surrounds the surface exposed to our eyes, as the atmosphere, but also that which is in contact with all the remaining surfaces of the body which are continuous with it through the larger foramina of the surface, such as the entire surface of the trachea, with which the air inhaled in breathing comes in contact; the entire surface of the alimentary canal, by which I mean the mouth, the œsophagus, the stomach, and the intestines; the entire surface of the bladder, and of the urethra; the entire surface which communicates with the uterus, especially in the years of puberty; the entire surface of all the excretory organs from the capillaries even to the orifices which discharge their contents into the ears, eyelids, nose, eyes, alimentary canal, bladder, urethra, uterus, and skin — a separate enumeration of which would show that many are truly external which are commonly considered internal, nay, even internal in the highest degree, and hence it follows:

    1. That worms and calculi are generated within our body and that most are formed in the external fluid.
    2. That many parts are essential to certain animals because they have them, not because the animal cannot exist without them.

A fluid which is in contact with these surfaces I call external because it communicates with the surrounding fluid by means of canals without intermediate capillary veins, that is, without cribration; and the result is that although the cavities containing the fluids mentioned may be closed at times, still whenever they are opened, they discharge all their retained fluid without a dividing membrane.

I call that fluid internal which does not communicate with the external fluid except through the intermediate strainers of the capillary veins, and therefore never discharges all its contents naturally into the external fluid without a dividing membrane.

The internal common fluid is that which is contained in the veins, arteries, and lymphatic ducts, at least in those ducts which connect the conglobate glands[1] and the veins. I call this fluid common because it is distributed over all parts of the body. Concerning another common fluid which resides in the nerve substance, I have nothing to say, because it is unknown.

The internal peculiar fluid is that fluid which is sent about in the capillary veins of the common fluid, and which varies with the diversity of places; for it is one thing in bloody parenchymata,[2] another in bloodless parenchymata, another around the muscle fibres,[3] another in the capsule of the ovum, another in the substance of the uterus, and still another in other places. For that belief accords with neither reason nor experience, which supposes that the ends of the veins and arteries terminate in the smallest possible particle of the body for the distribution of warmth and nourishment at that point.[4] But there are cavities everywhere, and the elements which have been secreted into these cavities from the blood are mingled with the fluid of that

[1] The lymphatic glands, as shown by Steno's treatise *De Glandulis Oris et Novis inde Prodeuntibus Salivae Vasis*, printed by Maar (Vol. I, p. 20, and note, p. 227).

[2] The name *parenchyma* was given by Erasistratus to the peculiar substance of the lungs, liver, kidneys, and spleen, on the theory that this substance, as distinguished from the flesh of the muscles, was formed from the blood which flowed from the veins and coagulated in the organs mentioned. See Galen, Περὶ τῆς τῶν Φαρμάκων Κράσεως καὶ Δυνάμεως, book XI, prooemium; and εἰς τὸ περὶ φύσεως ἀνθρώπου Βιβλίον Ἱπποκράτους Ὑπόμνημα πρῶτον, 4 (edition of Mewaldt, Helmreich, and Westenberger, Leipzig, 1914, p. 6).

[3] Steno's phrase is *circa fibras motrices*. This is defined in *Elementorum Myologiae Specimen*, Maar, *op. cit.*, Vol. II, p. 69: 'The *fibra motrix* is a certain bundle of very minute fibrillae closely joined longitudinally. . . . I call such a fibre *motrix* because it seems to me to be the true organ of motion in an animal. For the muscle is nothing except a collection of such fibres.'

[4] Steno refers to the theory commonly accepted before Harvey's demonstration of the circulation of the blood. This is the view expounded in Plato's *Timaeus* (79, 80), a work which, in a crude Latin translation, profoundly influenced the science of the Middle Ages.

P. 22. particular place, thereupon to be added to the solid parts, just as the particles worn from the solid parts flow back into the same cavities, to be restored to the blood again in order that with its help they may be carried back to the external fluid. The doctrine concerning the fluid of these cavities agrees in many ways with the teaching of the great Hippocrates [1] concerning air.

Although I may not be able to determine why different fluids are secreted in different places from the same blood, I hope, nevertheless, that a few things hold true for determining that question; since it is certain that such secretion does not depend upon the blood, but upon the places themselves. An examination of this matter is comprised under these three heads:

1. The consideration of the capillary veins of the internal common fluid; with which alone they concern themselves who attribute all things to cribration through the different pores — with whom I, too, was for a long time numbered.

2. The consideration of the internal peculiar fluid, with which alone they busy themselves who ascribe a special ferment to each part. Their belief may be partially true, although the name ferment rests upon a comparison taken [2] from too specific a process.

3. The consideration of the particular parts of a solid; and upon this they especially lay emphasis who acknowledge, by ascribing to each part its own form, that they recognize in it something peculiar to that part, but which is unknown to us, and which, according to the knowledge of matter which we have so far gained, can be nothing else than the porous surface of

P. 23. that solid, and the attenuated fluid permeating the pores.

I should wander too far afield if I were to apply the foregoing statements to the explanation of those things which take place in our body every day, and still they cannot otherwise be explained. It will suffice to have hinted here that the particles which separate from the external fluid in various ways are carried into the internal common fluid through the intervening

[1] The Περὶ Φυσῶν of Hippocrates assigns air in the body as the cause of all diseases, and different diseases as merely due to different organs thus affected; cf. the edition of Hippocrates by Littré, Vol. VI, pp. 92, 104–106, and Maar, *op. cit.*, Vol. II, p. 335.

[2] For *desumpta* in the Florentine edition of 1669 read *desumptae*.

cribration, and that after having been secreted from it likewise in various ways, and having been transmitted into the internal peculiar fluid by fresh cribration, they are added to the solid parts in the form of either fibres or parenchymata, according as they have been directed by the property of a given part, unknown to us, included in the consideration of the three foregoing statements.

If, now, we wish to reduce to definite classes the solids naturally contained within solids in the fashion indicated, we shall find some of them produced by accretion from the external fluid, which are due either to deposits, as the strata of the earth; or to incrustations, as the agate, onyx, chalcedony, eaglestone,[1] bezoar,[2] and so on; or to filaments, as the amian-

[1] The Latin word is *aëtites*. The 'eaglestone' is defined thus by the *New Oxford Dictionary*: "A hollow nodule or pebble of argillaceous oxide of iron, having a loose nucleus, which derived its name from being fabled to be found in the eagle's nest, and to which medicinal and magical properties were ascribed."

Some of these properties are mentioned by Damigeron, *de Lapidibus, Lapis Aëtites*:

'The aëtites is a very great safeguard of nature; God gave this stone to men as a protection to health. The eagle carries the stone to its nest from the uttermost parts of the earth for the sake of guarding its eggs. . . . The aëtites has a purple color and a very rough appearance, and has another stone within it, as if it were pregnant. It is useful to pregnant women, for when bound upon the left arm it prevents abortion. It is also very useful for accelerating parturition. For if taken from the woman's arm, and ground and placed upon her back, it will bring her immediate release. Furthermore, it will preserve the one who wears it, for it will make him sober and superior to all things; it will increase his wealth and spread about his good repute, and he will be most agreeable. . . . It is a remedy for insanity and unspeakable terrors, preserving the sufferers from dreaming and frequently falling. If you suspect that there is a poisoner in your home, put the stone in a relish and invite the suspect to dinner. If he is a poisoner, he will eat nothing, and if he ventures to swallow, he will not be able to do so. Such power has this stone. But if you remove the stone from the relish, the criminal will begin to eat and make merry. The wearing of the stone also greatly lessens anger incurred from powerful men. This stone is a sort of safeguard; the eagle uses it as a preventive against harm. For the eagle carries it from a never-failing river and puts it against the young in the nest to keep them from being harmed by another bird.' Abel, *Orphei Lithica, accedit Damigeron de Lapidibus* (Berlin, 1881), pp. 163, 164.

Similar statements are found in Pliny, *N. H.*, X. 12 (3); XXX. 130 (14). Val. Rose, in his study of the sources of Damigeron (*Hermes*, Vol. IX, 1875, pp. 471–491), cites an interesting parallel from Demosthenes, *Concerning Stones*, pp. 481, 482.

Palissy (cf. p. 208) ventures an independent opinion as to the origin of the aëtites: "Il y a beaucoup d'autres pierres qui sont formées selon le suiet qu'ils ont pris, comme quelques autres pierres que i'ay veuës que l'on nomme Pierre d'Aigle. Quelque chose que l'on en die, ie croy que ce n'est autre chose qu'un fruit lapifié, et ce qui iouë dedans est le noyau, qui estant amoindry quand on secouë ladite pierre, ledit noyau frappe des deux costez d'icelle." *Œuvres Complètes de Bernard Palissy*, by Paul-Antoine Cap, Paris, 1844, p. 284.

[2] An account of the medical history of the *Bezoar* is given by A. Laboulbène in *Diction-*

thus, feathery alum, different kinds of veins which I have observed in the fissures of rocks; or to dendrites, as those forms of plants which are seen in the chinks of stones, except that certain ramifications in an agate which I have seen, whose trunks rested on the surface of the outer lamella but whose branches spread throughout the substance of the inner lamella, are merely superficial; or to angular bodies, as the crystals of mountains, the angular bodies of iron and copper, cubes[1] of marcasites, diamonds, amethysts, and the like; or to replacements, as variegated marbles of every kind, granites, dendrites, petrified mollusks, crystalline substances, metallic plants, and many similar bodies filling the places of bodies which have been destroyed.

P. 24.

Other solids are produced by accretion from the internal fluid; and these are due either to simple replacements, as fat, the callus uniting broken bones, the cartilaginous substance joining severed tendons, the tissues which chiefly form the substance of the viscera, the medulla in both plants and animals; or to fibrous growths, as the fibrous parts of plants, the nerve fibres and muscle fibres in animals, also, which are all solid bodies and are naturally enclosed, for the most part, within solids.

If, therefore, every solid has had its accretions, at any rate, from a fluid, if bodies similar to one another in all respects were also produced in a similar way, and if of two contiguous solids that one first became hard which exhibits on its own surface the

---

naire *Encyclopédique des Sciences Médicales*, Tome Neuvième (Paris, 1868), pp. 221–225. Steno refers, of course, to the fossil, which is briefly alluded to in the work mentioned, p. 225: "Le bézoard fossile était composé de masses globuleuses de carbonate de chaux, réunies en couches concentriques."

A fuller description of the stone is quoted by Maar (*op. cit.*, Vol. II, p. 336) from the *Dictionnaire Raisonné Universel d'Histoire Naturelle*, by Valmont-Bomare, 3d ed., Lyon, 1791, Vol. II, p. 230:

"Une pierre arrondie, de couleur cendrée, composée de couches concentriques, friables, depuis la grosseur d'une aveline jusqu'à celle d'un œuf d'oie. Au centre de cette pierre est quelquefois un grain de sable, une petite coquille, ou un morceau de charbon de terre. Une de ces matieres a servi de noyau, de point d'appui, et venant à rouler sur des terres molles, à demi-trempées, elle s'est ainsi accrue par couches roulées comme une pelotte de rubans."

[1] For *ubi* of the Florentine edition read *cubi*, with Maar, *op. cit.*, Vol. II, p. 195. By "cubes of marcasite" Steno, in common with the older scientists, means pyrites. Marcasite has the same chemical composition as pyrites, FeS_2, being iron disulphide. But marcasite crystallizes in orthorhombic form, whereas pyrites crystallizes in the cubic system. See H. N. Stokes, *On Pyrite and Marcasite*, in *Bulletin of United States Geological Survey*, No. 186 (1901).

characteristics of the other's surface, it will be easy, granted a solid and the place in which it is, to affirm something definite about the place of its production. And this, indeed, is the general question *of a solid contained within a solid.*

P. 25. I pass to the more particular investigation of those solids dug from the earth which have given rise to many disputes; especially incrustations, deposits, angular bodies, the shells of marine animals, of mollusks, and the forms of plants. Under incrustations belong rocks of every kind consisting of layers, whose two surfaces are indeed parallel but not extended in the same plane. The place where incrustations are formed is the entire common boundary of fluid and solid; and the result is that the form of the layers or crusts corresponds to the form of the place, and it is easy to determine which of them hardened first, which last. For if the place was concave, the outer layers were formed first; if convex, the inner;[1] if the place was uneven because of various larger projections, the new layers were produced in the larger spaces when the narrower spaces had been filled with the formation of the first layers.

From this fact it is easy to account for all the differences of form which are seen in sections of similar rocks, whether they show the round veins of a tree cut transversely, or resemble the sinuous folds of serpents, or run along, curved in any other way, without law. Nor is it surprising that agates and other kinds of incrustations seem, so far as regards their outer surface, rough like ordinary stones,[2] since the outer surface of the outer layer imitates the roughness of the place. In torrents, however, incrustations of this kind are more frequently found outside of the place of their production, because the matter of the place has been scattered by a breaking up of the strata.

Concerning the manner in which particles of the layers
P. 26. which are to be added to a solid are separated from the fluid, the following at least is certain:

---

[1] The reference is doubtless to the formation of secretions in the first instance, and concretions in the second.

[2] The Florentine edition reads *saxis ignobilis instar asperos*; this is partially corrected in the Leyden edition of 1679 to *saxi*, etc. The correct reading *saxi ignobilis instar asperas* is given by Maar, *op. cit.*, Vol. II, p. 196.

1. That there is in it no place for buoyancy or gravity.

2. That the particles are added to surfaces of every kind, since surfaces smooth, rough, plane, curved, and consisting of several planes at different angles of inclination, are found overspread by the layers.

3. That movement of the fluid causes them no hindrance.

Whether the substance under consideration which flows from a solid, be different from that substance which moves the parts of the fluid, or whether something else is to be sought, I leave undecided.

Different kinds of layers in the same place can be caused either by a difference of the particles which withdraw from the fluid one after the other, as this same fluid is gradually disintegrated more and more, or from different fluids carried thither at different times. From this fact it follows that the same arrangement of layers sometimes recurs in the same place, and often evident traces revealing the entrance of new matter remain. But all the matter of the layers seems to be a finer substance emanating from the stones, as will further appear in the following.

THE STRATA OF THE EARTH

The strata of the earth are due to the deposits of a fluid:

1. Because the comminuted matter of the strata could not have been reduced to that form unless, having been mixed with some fluid and then falling from its own weight, it had been spread out by the movement of the same superincumbent fluid.

2. Because the larger bodies contained in these same strata obey, for the most part, the laws of gravity, not only with respect to the position of any substance by itself, but also with respect to the relative position of different bodies to each other.

3. Because the comminuted matter of the strata has so adjusted itself to the bodies contained in it that it has not only filled all the smallest cavities of the contained body, but has also expressed the smoothness and lustre of the body in that part of its own surface where it is in contact with the body, although the roughness of the comminuted matter by no means admits of similar smoothness and lustre.

Sediments, moreover, are formed so long as the contents in

a fluid fall to the bottom of their own weight, whether the said contents have been carried thither from some other where, or have been secreted gradually from the particles of the fluid, that too, either in the upper surface, or equally from all the particles of the fluid. Although a close relationship exists between crusts and sediments, they can nevertheless be distinguished easily because the upper surface of crusts is parallel to the lower surface, however rough[1] this may be from various larger projections, while the upper surface of sediments is parallel to the horizon, or deviates but slightly therefrom. So in rivers, the mineral layers, now green, now yellow, now reddish, do not remove the unevenness of a stony bottom, while a sediment of sand or clay makes all level; and it is due to this fact that in the formation of the different composite strata of the earth I have easily distinguished crusts from sediments.

Concerning the matter of the strata the following can be affirmed:

1. If all the particles in a stony stratum are seen to be of the same character, and fine, it can in no wise be denied that this stratum was produced at the time of the creation from a fluid which at that time covered all things; and Descartes[2] also accounts for the origin of the earth's strata in this way.

2. If in a certain stratum the fragments of another stratum, or the parts of animals and plants are found, it is certain that the said stratum must not be reckoned among the strata which settled down from the first fluid at the time of the creation.

3. If in a certain stratum we discover traces of salt of the sea, the remains of marine animals, the timbers of ships, and a substance similar to the bottom of the sea, it is certain that the sea was at one time in that place, whatever be the way it came there, whether by an overflow of its own or by the upheaval of mountains.

4. If in a certain stratum we find a great abundance of rush, grass, pine cones, trunks and branches of trees, and similar ob-

---

[1] *aspera*, Florentine edition, is an error for *asperae*.
[2] The reference is to Descartes, *Principia Philosophiae* (first edition Amsterdam, 1644), *Pars Quarta*, XXXII ff. See *Œuvres de Descartes, Publiées par Charles Adam et Paul Tannery*, Paris, Vol. VIII (1905), p. 220 ff.

jects, we rightly surmise that this matter was swept thither by the flooding of a river, or the inflowing of a torrent.

5. If in a certain stratum pieces of charcoal, ashes, pumice-stone, bitumen,[1] and calcined matter appear, it is certain that a fire occurred in the neighborhood of the fluid; the more so if the entire stratum is composed throughout of ash and charcoal, such as I have seen outside the city of Rome, where the material for burnt bricks is dug.

6. If the matter of all the strata in the same place be the same, it is certain that that fluid did not take in fluids of a different character flowing in from different places at different times.

7. If in the same place the matter of the strata be different, either fluids of a different kind streamed in thither from different places at different times (whether a change of winds or an unusually violent downpour of rains in certain localities be the cause) or the matter in the same sediment was of varying gravity, so that first the heavier particles, then the lighter, sought the bottom. And a succession of storms might have given rise to this diversity, especially in places where a like diversity of soils is seen.

8. If within certain earthy strata stony beds are found, it is certain either that a spring of petrifying waters existed in the neighborhood of that place, or that occasionally eruptions of subterranean vapors occurred, or that the fluid, leaving the sediment which had been deposited, again returned when the upper crust had become hardened by the sun's heat.

Concerning the position of strata, the following can be considered as certain:

1. At the time when a given stratum was being formed, there was beneath it another substance which prevented the further descent of the comminuted matter; and so at the time when the lowest stratum was being formed either another solid substance was beneath it, or if some fluid existed there, then it was not only of a different character from the upper fluid, but also heavier than the solid sediment[2] of the upper fluid.

---

[1] The inclusion of bitumen in the list indicates that Steno was ignorant of its true nature as an organic compound.

[2] *sedimenta*, Florentine edition, is an error for *sedimento*.

2. At the time when one of the upper strata was being formed, the lower stratum had already gained the consistency of a solid.

3. At the time when any given stratum was being formed it was either encompassed on its sides by another solid substance, or it covered the entire spherical surface of the earth. Hence it follows that in whatever place the bared sides of the strata are seen, either a continuation of the same strata must be sought, or another solid substance must be found which kept the matter of the strata from dispersion.

4. At the time when any given stratum was being formed, all the matter resting upon it was fluid, and, therefore, at the time when the lowest stratum was being formed, none of the upper strata existed.

As regards form, it is certain that at the time when any given stratum was being produced its lower surface, as also its lateral surfaces, corresponded to the surfaces of the lower substance and lateral substances, but that the upper surface was parallel to the horizon, so far as possible; and that all strata, therefore, except the lowest, were bounded by two planes parallel to the horizon. Hence it follows that strata either perpendicular to the horizon or inclined toward it, were at one time parallel to the horizon.

Moreover, the changed position of strata and their exposed sides, such as are seen to-day in many places, do not contradict my statements; since in the neighborhood of those places evident traces of fires and waters are to be found. For just as water disintegrating earthy material carries it down sloping places not only on the surface of the earth but also in the earth's cavities; so fire, breaking up whatever solids oppose it, not only drives out their lighter particles but also sometimes hurls forth their heaviest weights; and the result is that on the surface of the earth are formed steeps, channels, and hollows, while in the bowels of the earth subterranean passages and caverns are produced.

By reason of these causes the earth's strata can change position in two ways:

The first process is the violent thrusting up of the strata, whether this be due to a sudden burning of subterranean gases, or be brought about through the violent explosion of air due to other great downfalls near by. This thrusting up of the strata is followed by a scattering of the earthy matter as dust and the breaking up of rocky matter into lapilli and rough fragments.

The second process is the spontaneous slipping or downfall of the upper strata after they have begun to form cracks, in consequence of the withdrawal of the underlying substance, or foundation. Hence by reason of the diversity of the cavities and cracks the broken strata assume different positions; while some remain parallel to the horizon, others become perpendicular to it, many form oblique angles with it, and not a few are twisted into curves because their substance is tenacious. This change can take place either in all the strata overlying a cavity, or in certain lower strata only, the upper strata being left unbroken.

The altered position of the strata affords an easy explanation of a variety of matters otherwise obscure. Herein may be found a reason for that unevenness in the surface of the earth which furnishes occasion for so many controversies; an unevenness manifest in mountains, valleys, elevated bodies of water, elevated plains, and low plains. But passing over the rest, I shall now treat briefly certain points concerning mountains.

THE ORIGIN OF MOUNTAINS

That alteration in the position of strata is the chief cause of mountain formation is clear from the fact that in any given range of mountains there may be seen:

1. Large level spaces on the summits of some mountains.
2. Many strata parallel to the horizon.
3. Various strata on the sides of the mountains inclined at different angles to the horizon.
4. Broken strata on the opposite sides of hills, showing absolute agreement in form and material.
5. Exposed edges of strata.
6. Fragments of broken strata at the foot of the same range, partly piled into hills, and partly scattered over the adjoining country.

7. Either in the rock of the mountains themselves, or in their neighborhood, very clear traces of subterranean fire; just as many springs are found around hills which are made up of strata of earth. And here it must be observed in passing that the hills which are formed of earthy strata, for the most part, have as their foundation larger fragments of stony strata; these in many places keep the earthy strata placed upon them from being swept away by the current of neighboring rivers and torrents. Further, they often protect entire regions against the violence of the ocean, as the extended reefs of Brazil[1] and exposed craggy shores everywhere bear witness.

Mountains can also be formed in other ways, as by the eruption of fires which belch forth ashes and stones together with sulphur and bitumen; and also by the violence of rains and torrents, whereby the stony strata, which have already become rent apart by the alternations of heat and cold, are tumbled headlong, while the earthy strata, forming cracks under great blasts of heat, are broken up into various parts. And from this it is clear that the chief classes of mountains and hills are two: first, of those which consist of strata; of these there are two kinds, since in some, strata of rock prevail, in others, strata of earth. The second class is composed of mountains which rise without order or arrangement from fragments of strata and from parts, further, which have been worn away.

Hence it could be easily shown:

1. That all present mountains did not exist from the beginning of things.

P. 34.   2. That there is no growing[2] of mountains.

---

[1] Steno's information regarding Brazil was probably gained from a book called *Historia Naturalis Brasiliae*, Amsterdam, 1648. The volume contains Piso's *De Medicina Brasiliensi Libri Quatuor*, and George Musgrave's *Historiae Rerum Naturalium Brasiliae Libri Octo*. No doubt Casper Barlaeus's *Rerum per Octennium in Brasilia sub Praefectura Mauritii Nasovii Historia* (Amsterdam, 1647) was also known to him.

Willem Piso (1611–1678) was a member of the Brazilian expedition of Count Jan Maurits from 1636 to 1644. Steno had known Piso in Leyden and in 1664 addressed to him the letter on the Anatomy of the Ray (*De Anatome Rajae Epistola*) printed by Maar, *Opera Philosophica*, Vol. I, pp. 193–207. Robert Boyle refers frequently to Piso's *History of Brazil*.

[2] Steno's word is *vegetatio*, which suggests the growth of an organism; but he does not hesitate to use *crescere* of inorganic accretions. The passage quoted by Maar, *op. cit.*, Vol. II, p. 338, from Fabronius (*Vitae Italorum* (p. 202), Vol. III, p. 72), is singularly apposite. In

3. That the rocks or mountains have nothing in common with the bones of animals except a certain resemblance in hardness, since they agree in neither matter nor manner of production, nor in composition, nor in function, if one may be

1657 Montanari and Boni, master of the mint in Vienna, journeyed to Stiermark, Bohemia, and Hungary to examine the mines. I translate:

'They also investigated whether metals grow in the same manner as plants do, that is by means of a circulating sap of the earth. They thought they knew, from the surest proofs, that metals do indeed grow (*crescere*) — iron rather rapidly and gold more slowly. But how this took place they were unable to decide, although Montanari inclined to believe that the growth (*maturitatem*) was caused by accretion (*fermentationi*). . . . And he made fun of the levity and weakness of those who believed the testimony of George Agricola that gnomes flit and wander about the mines, by whom the workmen are often disturbed.'

Agricola (1494–1555), scientist though he was, fully believed in gnomes. Compare *De Re Metallica*, Book VI, p. 217, edition of Hoover (London, 1912):

"In some of our mines, however, though in very few, there are other pernicious pests. These are demons of ferocious aspect, about which I have spoken in my book *De Animantibus Subterraneis* (the last paragraph). Demons of this kind are expelled and put to flight by prayer and fasting."

Agricola's credulity, however, did not extend to a belief in the "growth" of mountains. On the contrary, he was perhaps the first to recognize clearly the fundamental agencies of mountain sculpture, as appears from *De Ortu et Causis Subterraneorum*, Book II (*De Re Metallica*, edition of Hoover, pp. 595, 596):

"Hills and mountains are produced by two forces, one of which is the power of water, and the other the strength of the wind. There are three forces which loosen and demolish the mountains, for in this case, to the power of the water and the strength of the wind we must add the fire in the interior of the earth. Now we can plainly see that a great abundance of water produces mountains, for the torrents first of all wash out the soft earth, next carry away the harder earth, and then roll down the rocks, and thus in a few years they excavate the plains or slopes to a considerable depth; this may be noticed in mountainous regions even by unskilled observers. By such excavation to a great depth through many ages, there rises an immense eminence on each side. When an eminence has thus arisen, the earth rolls down loosened by constant rain and split away by frost, and the rocks, unless they are exceedingly firm, since their seams are similarly softened by the damp, roll down into the excavations below. This continues until the steep eminence is changed into a slope. Each side of the excavation is said to be a mountain, just as the bottom is called a valley.

"Streams, moreover, and to a far greater extent rivers, effect the same results by their rushing and washing; for this reason they are frequently seen flowing either way between very high mountains which they have created, or close by the shore which borders them. . . . Nor did the hollow places which now contain the seas all formerly exist, nor yet the mountains which check and break their advance, but in many parts there was a level plain, until the force of winds let loose upon it a tumultuous sea and a scathing tide. By a similar process the impact of water entirely overthrows and flattens out hills and mountains. But these changes of local conditions, numerous and important as they are, are not noticed by the common people to be taking place at the very moment when they are happening, because, through their antiquity, the time, place, and manner in which they began is far prior to human memory.

"The wind produces hills and mountains in two ways: either when set loose and free from bonds, it violently moves and agitates the sand; or else when, after having been driven into the hidden recesses of the earth by cold, as into a prison, it struggles with a great effort to burst out. For hills and mountains are created in hot countries, whether they are situated by

permitted to affirm aught about a subject otherwise so little known as are the functions of things.

4. That the extension of crests of mountains, or chains, as some prefer to call them, along the lines of certain definite zones of the earth, accords with neither reason nor experience.[1]

5. That mountains can be overthrown, and fields carried over from one side of a high road across to the other; that peaks of mountains can be raised and lowered, that the earth can be opened and closed again, and that other things of this kind occur which those who in their reading of history wish to escape the name of credulous, consider myths.[2]

PASSAGE-WAYS FOR THINGS ISSUING FROM THE EARTH

The same alteration in the position of strata affords a passage-way for things issuing from the earth, such as:

1. Waters, which are shut up in mountain caves away from the air, gushing forth on the mountains, whether those waters come from subterranean reservoirs, or have been condensed in a place away from the upper air and then ejected. And this I believe to be very common, since in many caverns I have observed that everything both above and below was solid though the water nevertheless trickled there abundantly.

2. Winds breaking forth from mountains, whether those

the sea coasts or in districts remote from the sea, by the force of winds; these no longer held in check by the valleys, but set free, heap up the sand and dust, which they gather from all sides, to one spot, and a mass arises and grows together. If time and space allow, it grows together and hardens, but if it be not allowed (and in truth this is more often the case), the same force again scatters the sand far and wide. . . .

"Then, on the other hand, an earthquake either rends and tears away part of a mountain, or engulfs and devours the whole mountain in some fearful chasm. In this way it is recorded the Cybotus was destroyed, and it is believed that within the memory of man an island under the rule of Denmark disappeared. Historians tell us that Taygetus suffered a loss in this way, and that Therasia was swallowed up with the island of Thera. Thus it is clear that water and the powerful winds produce mountains, and also scatter and destroy them. Fire only consumes them, and does not produce at all, for part of the mountains — usually the inner part — takes fire."

[1] Steno is not referring to mountain-chains in the modern sense of the term; he is rejecting Kircher's theory of chains running from north to south and east to west over the entire surface of the earth. This is set forth in *Mundus Subterraneus*, Amstelodami (1665), Vol. I, c. ix, p. 68 ff. Cf. Maar, *op. cit.*, Vol. II, p. 337.

[2] For a modern exposition of Tuscan earth features, see Murchison, *Geological Structure of the Alps*, in *Quarterly Journal of the Geological Society*, vol. 5 (1849), pp. 157–312, especially pp. 263–308.

winds be air expanded by heat or whether different fluids of the air made violent by collision produced them.

P. 35.

3. Ill-smelling exhalations, fiery or frigid ebullitions, and so on. And there is no longer any doubt of the fact that cold and dry places boil up without any trace of heat whenever water is poured upon them; that a hot spring issues by the side of a very cold spring; that in consequence of an earthquake a hot spring may be turned into a cold spring, and rivers change their course; that valleys shut in on all sides discharge their gathered rain water into lower places; that rivers gliding underneath the earth's surface are in places returned to the light of day; that in laying foundations architects sometimes lose all their labor when they encounter a quicksand, as it is called;[1] that in certain places on digging wells water is at first found near the surface of the ground, then after digging to the depth of several yards new waters are discovered which at first, on the opening of a passage-way, leap forth beyond the height of the water already found; that whole fields with trees and buildings sink gradually, or are engulfed suddenly, and hence vast lakes now exist where once stood cities; that a plain is a source of danger to its inhabitants from catastrophes of this kind unless they have made themselves sure about its foundation of rock; that abysses emitting a deadly gas are sometimes found which are again stopped up when a number of bodies have been cast into them.

THE ORIGIN OF VARIEGATED STONES AND THE REPOSITORIES OF MINERALS

The same alteration in the position of strata has given rise to variegated stones of every kind, and at the same time afforded a repository for most minerals, whether the deposition took place in the cracks of the strata, or in those fissures which were, in respect to matter, dry but not yet hard, either between the layers or in their clefts; or in the interstices between the upper and the lower strata after the falling of lower strata; or in the places left empty by the decomposition of bodies therein contained. Whence it can be shown:

P. 36.

1. That on the very slightest foundation, nay, apparently on

[1] *arena viva,* 'living sand,' is Steno's phrase.

no foundation, have been based those minute and all but inconceivable subdivisions of veins made use of by diggers of minerals; and that divination for the abundance of metal by means of roots and branches is, in consequence, as doubtful as is the ridiculous belief of certain Chinese concerning the head and tail of the dragon which they employ in finding a favorable place of burial in the mountains.[1]

2. That most of the minerals for which man's labor is spent did not exist at the beginning of things.

3. That in the investigation of rocks many things can be disclosed which are attempted in vain in the study of the minerals themselves, seeing that it is more than probable that all those minerals which fill either the clefts or expanded spaces of rocks had as their matter the vapor forced from the rocks themselves, whether the deposition took place before the strata changed their position, as I believe happened in the mountains of Peru,[2] or when the strata had already changed their position; and that a new metal can therefore form in the place of an exhausted one, as is believed rather than known concerning the mining of iron among the people of Elba, for the miner's tools and the idols which have been found there were surrounded not with iron but with earth.[3]

And these things concerning the strata of the earth I thought ought to be investigated the more carefully, not only because the strata themselves are solids naturally enclosed within solids but also because in them are contained almost all those bodies which gave rise to the question propounded.

---

[1] This practice is mentioned by Kircher, *China Illustrata* (Amstelodami, 1667), p. 135, who quotes Trigautius, *De Christiana Expeditione apud Sinas Suscepta* (Augustae Vind., 1615), lib. I, ix, p. 95. 'One characteristic of the Chinese can be mentioned. In seeking a spot for building private and public structures, or for burying their dead, they examine the spot with the head, tail, and feet of various dragons which are supposed to live beneath it; and they believe that all their adversity and prosperity depend upon the dragons.' Maar, *op. cit.*, Vol. II, p. 337.

What is perhaps the first published description of the divining rod and its use in finding minerals or water, is given by Agricola, *De Re Metallica*, II, edition of Hoover, pp. 38-42. See also Robert Boyle, edition of Shaw, Vol. I, pp. 172, 173.

[2] The mineral deposits in the Peruvian mountains were familiar to Steno from de Acosta's *Historia Natural y Moral de Las Indias* (Seuilla, 1590), iv, iv-v, and from de Rosnel's *Le Mercure Indien, ou Le Tresor des Indes* (Paris, 1667), Première Partie, Livre Premier, I-III. Maar, *op. cit.*, Vol. II, p. 337.

[3] The belief that iron would "grow" or replace itself in process of time, probably arose from finding limonite upon the tools mentioned in the text. See above, p. 232, note 2.

CONCERNING THE CRYSTAL

As regards the formation of crystal, I would not venture to declare in what manner its first shape is produced; this at any rate is beyond dispute, that the things which it has been my lot to read in other writers concerning this subject are not to the point; for neither irradiations, nor a shape of the particles resembling the shape of the whole, nor the perfection of the hexagonal form[1] and the assembling of the parts about a common centre, nor other things of this kind, accord with fact; as will be clear from various propositions which I shall bring forward, proved elsewhere by conclusive experiments. But that no room may be left for doubt, it is well to explain beforehand the terms which I employ in naming the parts of a crystal.

A crystal consists of two hexagonal pyramids and an intermediate prism likewise hexagonal. I call those angles the *terminal solid angles* which form the apexes of the pyramids, but those angles the *intermediate solid angles* which are formed by the union of the pyramids with the prism. In the same way I call the planes of the pyramids *terminal planes*, and the planes of the prism the *intermediate planes*. The *plane of the base* is the section perpendicular to all the intermediate planes; a *plane of the axis* is a section in which lies the axis of the crystal, which consists of the axes of the pyramids and the axis of the prism.

The place where the first hardening of a crystal begins, whether it be between a fluid and a fluid, or between a fluid and a solid, or even in a fluid itself, may remain in doubt; but the place in which the crystal grows after it has already begun to form, is a solid in that part where the crystal is supported on it, whether the place be a stone or another crystal already formed. The remaining portion is fluid, if you except the obstructions which can present themselves to it from the unevenness of the rock or even from other crystals already formed. I would not venture to affirm whether the surrounding fluid is aqueous; and it matters not what is said about the

---

[1] By crystal Steno meant rock crystal, which is the mineral quartz and has a hexagonal form.

water enclosed within crystals, since it is certain that air together with water is contained therein, and that many crystals are found which enclose air alone. But if the crystal had indeed grown in an aqueous fluid, all the spaces enclosed on every side would be filled with water,[1] since it is an undisputed fact that water kept in that way never vanishes in any number of centuries.

The cavities of the rocks, formed in different ways, as has been said above, afford this place for crystals, and the fact that entire hills consist of earthy substance packed with crystals, is no disproof, seeing that in the vicinity of the same hills are found mountains of rock suited to the formation of crystals.

P. 39. And in those hills of earthy matter, large unburied rocks are pulled out which have been rent from neighboring mountains, and some of these show fissures filled with the material of marble, precisely as the fissures of strata in mountains of rock are filled. The same cause, moreover, which hurled upon the hills the fragments of strata rent from the neighboring mountains, can likewise have sown over them broadcast the crystals which had been shaken out from cavities of the same strata.

The following propositions will show what can be determined concerning the place of the crystal to which new crystalline matter is being added:

I

A crystal grows while new crystalline matter is being added to the external planes of the crystal already formed. No room at all is here left for the belief of those who affirm that crystals grow, plantlike, by nourishment, and that they draw their nourishment on the side where they are attached to the matrix, and that the particles thus received from the fluid of the rock, and transmitted into the fluid of the crystal, are inwardly added to the particles of the crystal.

II

This new crystalline matter is not added to all the planes but, for the most part, to the planes of the apex only, or to the terminal planes, with the result:

[1] This would not necessarily follow.

1. That the intermediate planes, or the quadrilateral planes, are formed by the bases of the terminal planes, and hence the intermediate planes are larger in some crystals, smaller in others, and wholly wanting in still others.

2. That the intermediate planes are almost always striated, while the terminal planes retain traces of the matter added to them.

III

P. 40. The crystalline matter is not added to all the terminal planes at the same time, nor in the same amount. Hence it comes to pass:

1. That the axis of the pyramids does not always continue the same straight line with the axis of the prism.

2. That the terminal faces are rarely of a size, whence follows an inequality of the intermediate planes.

3. That the terminal faces are not always triangular, just as all the intermediate planes are not always quadrilateral.

4. That the terminal solid angle is broken up into several solid angles, this being the case frequently also with the solid intermediate angles.

IV

An entire plane is not always covered by crystalline matter, but exposed places are left sometimes toward the angles, sometimes toward the sides, and sometimes in the centre of the plane. Hence it happens:

1. That the same plane, commonly so-called, does not have all its parts located in the same plane, but in different planes extending above it in different ways.

2. That a plane, commonly so-called, in many places is seen to be not a plane but a protuberance.

3. That in the intermediate planes inequalities rise like the steps of stairs.

The crystalline matter added to planes upon the same planes is spread out by the enveloping fluid, and gradually hardens, with the result:

1. That the surface of the crystal comes forth the smoother the more slowly the added matter has hardened, and is left

wholly rough if the matter has hardened before it has spread sufficiently.

P. 41.   2. That the manner in which the crystalline matter is added to the crystal can be distinguished, since where it has hardened suddenly, it reveals a surface full of small elevations like variolar postules, as it were, just as small drops of oily fluid are wont to float upon an aqueous fluid; sometimes it shows also trilateral and depressed pyramids, if it has hardened somewhat more slowly. The tortuous fringes of the descending matter show now the place to which the fluid matter was being added, now the place toward which it was being advanced, now the arrangement of the matter added, that is, which came first, and which last. And in this way certain roughnesses always appear in the crystals of mountains, nor have I ever seen a crystal whose still unbroken surfaces possess the lustre which the rent sides of the same crystal show after it has been broken, however prolix writers on subjects relating to nature become in extolling the lustre of the crystal which is extracted from the mountains.

3. That certain intruding solid bodies are enclosed within the crystal itself, as if they had been coated with a sort of glue, in case the crystals did not yet present a hardened surface.

4. That the crystalline matter sometimes seems to flow down over neighboring planes.

5. That when certain small areas on those planes have been left without added crystalline matter, new crystalline matter approaches, and, spreading over the areas, forms cavities there, sometimes producing several layers; sometimes, again, enclosing a part of the external fluid, which in some instances is nothing but air, in others water and air.

P. 42.   The external fluid receives crystalline matter from the substance of the harder stratum, with the result:

1. That rocks of a different kind, emitting a different fluid, give rise to crystals of a different hue.

2. That in the same place sometimes the first, sometimes the last, crystals are the darker: but in the same crystal the parts first hardened are sometimes darker than the parts last hardened.

3. That when oysters, mussels, and other bodies have decomposed within the earth, the cavities are filled with crystals.

The movement of the crystalline matter toward a point where the planes of the crystal already formed are fixed, does not arise from some common cause of motion in the surrounding fluid, but varies in any given crystal; so that in reality it depends upon the movement of the tenuous fluid flowing from the crystal already formed, and the result is:

1. That in the same place crystalline matter is added to planes which face the horizon from different angles.

2. That in the same fluid crystals of different shapes are produced. Whether the fluid is that by the aid of which refraction is caused, or there is still some fluid different from it, I leave to wiser minds to study.

That the efficiency of a penetrating fluid is certainly great, is illustrated by the row of iron filings which rise about the poles[1] of a magnet, not only when the filings are in direct contact with the magnet but also when they are separated from the magnet by an intervening sheet of paper. When, for example, the magnet is moved in various ways below the paper, while one end remains at rest, filings of this kind describe on the paper all the arcs which can be drawn within a hemisphere. Now all advance from place to place like armed soldiers; now, deflected by the approach of another magnet, they form an arch just as if the individual parts of the filings had been glued together and had united into a solid body.[2]

---

[1] *Poros* in the original edition is an obvious error for *polos*.

[2] This experiment with the magnet was, no doubt, a scholastic commonplace. A curious analogue to Steno's illustration may be found in Robert Boyle's *The Effects of Languid Motion Consider'd* (edition of P. Shaw, London, 1725), Vol. I, pp. 477, 478:

"The load-stone is acknowledged to act by the emission of insensible particles. For tho' iron and steel be solid bodies, and magnetic effluvia corpuscles so very minute, as readily to get in at the pores even of glass itself; yet entring the steel in swarms, they may operate so violently on it, as to attract above fifty times the weight of the magnet. For to these I rather ascribe magnetical attraction and suspension, than to the pressure of the ambient air; because I have found on trial, that such a pressure is not absolutely necessary to magnetical operations.

"And farther, as to the power of magnetical effluvia upon iron, I took filings of iron fresh made, that the virtue might not be diminished by rust, and having laid them in a little heap upon paper held level, I applied to the lower side of it, just beneath the heap, the pole of a vigorous load-stone, whose emissions diffusing themselves thro' the metal, manifestly alter'd

In a similar way I should suppose that by the help of a permeating fluid those minute drops mutually cohere which have formed in a receiver from the material forced out of a retort. At first they cling together on the inside of the upper part of the receiver, but later, when a number of drops have come together in the upper part of the receiver, they fall down and form globular masses which sometimes cling, with their extremities, to the sides of the receiver, and sometimes join other filaments. Filaments of this kind, which I have sometimes noticed in the humor of the eye, I should believe to consist of globular masses and to have been formed in a similar way, and so, too, should I believe filaments and branches to have been produced in the fluid by accretion from without.

But however the case may stand concerning these things, in the growth of a crystal, we must take into account two movements: one, the movement whereby it is brought to pass that crystalline matter is added to certain places of the crystal and not to others — a movement which I fancy must be attributed to the attenuated permeating fluid, and is to be illustrated by the example of the magnet which I have given; the other the movement whereby the new crystalline matter added to the crystal is spread forth over the plane — and this movement must be derived from the surrounding fluid, just as, when the iron filings have risen up above the magnet, through the movement of the air, whatever is struck off from one filing is added to another. To this movement of the surrounding fluid I should attribute the fact that not only in a crystal, but also in many other angular bodies, any given opposite planes are parallel.

P. 44.

From the arguments presented it might be possible to prove that the efficient cause of the crystal is not extreme cold; that it is not ashes only, burned out by the force of fire, which turn into glass; that the force of fire alone is not the producer of glass; that not all crystals were produced in the beginning of things, but that they are even now being produced from day to day; that it is not a task beyond man's power to disclose the

their appearance, and produced many erect aggregates of filings, placed one above another, like little needles: and as these needles stood erected upon the flat paper, so they would run to and fro, according as the load-stone, which was held underneath, moved one way or the other; and as soon as that was taken away, all this little stand of pikes would fall again into a confused heap."

formation of glass without the agency of fire, provided one undertake a careful analysis of the rocks in whose cavities the best crystals are formed. For it is certain that, just as a crystal has formed from a fluid, so that same crystal can be dissolved into a fluid, provided one know how to imitate the real menstruum [1] of Nature. And it is no disproof that certain solid bodies, when once the dissolving fluid, or their menstruum, has been taken away, can be no further disintegrated by the same or a similar solvent; for this occurs in bodies from which the entire menstruum is freed by the force of fire. But the crystal, and all angular bodies which form in the midst of a solvent fluid, or in the midst of a menstruum, can never come out so pure but that some particles of the menstruum are left within the particles of the angular body. And upon this fact depends the main cause of variation whereby crystal differs from glass not

P 45. only in refraction but also in other properties, since in glass no parts of the dissolving fluid are present, inasmuch as they have been driven forth by the violence of fire. For the fluid, in which the crystal is formed, bears the same relation to the crystal that ordinary water bears to salts; this could easily be proved by setting forth the characteristics which the formation of salts holds in common with the formation of crystal.

But since I should be wandering too far from my subject if I should allude to all these things here, I shall mention but one example, which seemed exceedingly beautiful to me. In various places within the same stone the receding layers were filled with crystals, of which some were watery, others very clear, some white, many amethystine, mingled together without any blending of hues; exactly as experiments with salts made here show that vitriol and alum, dissolved in the same water, after a part of the water has been taken up, have each formed by themselves without any mixing of parts.

[1] The term *menstruum*, used by Steno, was commonly employed by the alchemists and physicists to denote a solvent fluid. Compare Littré, *Dictionaire de la Langue Française, s.v. menstrue*: "Terme de chimie. Liqueur propre à dissoudre les corps solides. L'eau régale est le menstrue de l'or *(aqua regia)*. On dit aujourd'hui de préférence dissolvant."

Excellent examples of this usage can be found in Robert Boyle's works; compare, *e.g.*, *Experiments and Observations upon Colours* (edition of Shaw, Vol. II, p. 96): "That gold, dissolv'd in *Aqua regia*, communicates its own colour to the menstruum, is a common observation; but the solutions of mercury, in *Aqua fortis*, are not generally observ'd to give any notable tincture to the menstruum. See also *New Oxford Dictionary, s.v. menstruum*.

ANGULAR BODIES OF IRON

The angular bodies of iron which it has hitherto been my fortune to see[1] reduce to three classes. Of these the first is plane and, being thicker in the middle, gradually grows thinner towards the margins, where it terminates in an edge sharp on every side; the second is bounded by twelve planes, the third by twenty-four planes. Sometimes an angular body of the second class is bounded by six planes, resembling two trilateral pyramids so joined along the bases that the angles of one base bisect the sides of the other.

P. 46.

The second and third classes of angular bodies of iron agree with crystals in the following particulars:

1. In the place of production; since the place where iron is formed is partly solid, partly fluid, and is a cavity in the rock.

2. As regards the place to which matter is added; since in iron also it is added not to all the planes, but to some only, and not always to the whole of these, nor always at the same time, but now to one, now to another; now towards the margins, and now towards the middle.

3. As regards the place from which the iron matter comes, since this matter, also, seems to have flowed forth from the pores of a more solid body.

4. As regards the manner in which the same matter is directed toward the solid by the help of the permeating fluid, and is spread forth and smoothed out upon the plane by the movement of the surrounding fluid.

Iron and crystal differ in matter and form, because the matter of the crystal is translucent, while the matter of iron is opaque. The form of the crystal is bounded by eighteen planes, of which the twelve terminal are brilliant, while the six intermediate are striated. In the second class of iron, however, twelve planes may be counted, of which six are terminal and striated, the other six intermediate and brilliant; and in the third class of iron twenty-four planes may be counted, of which the six terminal are striated, and the intermediate eighteen brilliant. Between the terminal striated planes there sometimes lie six other

P. 47.

---

[1] Steno refers to crystals of hematite from the mines on the island of Elba.

glistening planes resembling the truncated sides of triangular pyramids.

It seemed to me worthy of notice that by truncating a cube at the very extremity the entire number of planes in the third kind of angular bodies of iron can be shown; for it has six pentagonal planes which coincide with the planes of the cube, and which, at the four angles,[1] bisect the individual sides of the cube's planes. All the remaining planes are found at the cube's angles when they are truncated in a certain way.

In the angular bodies of iron there is also another thing equally surprising. In the second class of angular bodies of iron the terminal planes, which are striated and five-sided, are in process of time changed to three sides, while intermediate planes, which are three-sided and brilliant, pass into five-sided with two right angles adjacent to each other.[2] Between two five-sided planes, however, where their right angles are adjacent, a pair of triangles, or two three-sided planes, are formed, likewise brilliant, whose bases coincide with the perpendicular side of the five-sided planes; so that the second class of iron is thus changed into the third.

That in this same way a body of twenty-four planes is formed from a body of twelve, I am convinced for the following

P. 48. reasons: (1) Because in the same mass of iron bodies almost all the thinner bodies have only twelve planes, while the thicker ones have twenty-four. (2) Because in certain bodies of twelve planes are seen the beginnings of triangular planes which are accessory and which, if continued, form a body of twenty-four planes.

In triangular planes I have sometimes noticed a smoothness so perfect that not the slightest unevenness was apparent to the eye, — something which it has never yet been my lot to see in any crystal.[3] In other instances I have seen smaller curved planes piled above larger, of which the higher were, for the most part, nearest the triangular apex, so that one may therefore question whether the five-sided planes are not formed by

---

[1] The polyhedral angles. Steno is apparently referring to the relation of the rhombohedron to the cube.

[2] Steno evidently thought that the various modifications of hematite resulted from an evolution in time of new crystal forms.

[3] Quartz or rock crystal.

the bases of triangular planes, since traces of striæ appear in them parallel to the bases.

That in the case of copper, angular bodies are formed in the same way that has been mentioned in the case of the crystal and iron, is inferred from those bits of copper which you[1] preserve among the curiosities of nature,[2] but since the abundance of the matter has filled all the interstices of the bodies, it is difficult to ascertain the original form of the bodies. And precisely the same is true of the angular bodies of silver sent to you from Germany.

CONCERNING THE DIAMOND

Concerning the diamond, the same thing is inferred touching the place and manner of production which is inferred from the crystal, namely:

1. That diamonds have been produced in a fluid enclosed in the cavities of rocks, although a distinguished writer on India attempts to prove that diamonds are again produced in a certain period of years, in the earth from which they have once been dug.[3]

2. That they have been produced from a fluid by the accretion of diamond matter.

3. That in their production the workings of both the attenuated permeating fluid and the surrounding fluid must be taken into account.

For the rest, as regards the form of diamonds, it varies greatly, since some are bounded by eight planes, others by nine,[4] others by eighteen, others by twenty-four planes; and among these most of the planes were striated, while some were also smooth. Although some diamonds might be angular, they nevertheless could have some surfaces curved rather than plane.

---

[1] Ferdinand II. See p. 205.

[2] Steno refers to the collection in the Pitti Palace. See p. 182.

[3] Maar (*op. cit.*, Vol. II, p. 338) observes that Steno may have had in mind P. de Rosnel's *Le Mercure Indien* (Paris, 1667), *Seconde parte, livre premier*, Chapter II, p. 12: "Monardes en son livre . . . remarque que les grands diamants prennent d'ordinaire leur naissance de la partie inferieure de la mine, et que les petits prennent la leur de la partie superieure."

[4] Not a regular form of diamond but doubtless due to the disappearance of certain faces.

CONCERNING MARCASITES

The substance of marcasites assumes divers forms, for sometimes it incrusts the surface of a place, sometimes it is condensed into bodies of many planes, sometimes it forms rectangular parallelopipeds[1] which, after the usual mode of speech, we shall call cubes, although regularity of all the planes is found in but a few.

Since I have had the opportunity to note various matters concerning the cubes of marcasites, both the cubes themselves and the place where they are found, I shall speak concerning those matters only; but their formation, nevertheless, differs from the formation of a crystal:

1. In time; since the cubes of marcasites were formed before the formation of the strata in which they are contained, whereas crystals hardened after the formation of the strata.

2. In place of production; for a crystal, at least while it was forming, was resting upon a solid body and so was contained partly in a solid place, partly in a fluid, while the cubes of the marcasites seem to have formed between two fluids, since there are no traces, even in the larger cubes, of cohesion with another body; although small cubes are frequently found which, while growing, adhere to one another in the surface of the fluid. Moreover we are taught by the weighty proofs of the great Galileo[2] that heavier substances of this kind can cling together on the surface of a fluid while one of their surfaces is in immediate contact with an overlying and lighter fluid of another kind. That one of the fluids referred to was aqueous, is shown by the matter of the stratum which results from the same fluid.

3. In the manner and place of accretion; for the matter of the marcasite is added to all the planes of the cubes in a manner different from that which we have indicated in the case of crystals. This fact is clearly shown by the uniformity of all the surfaces of the cubes which I have myself cut from rocks; all the planes of these had striæ parallel to two sides, in such

---

[1] For Steno's use of the word *marcasites*, see p. 225, note 1.

[2] The treatise of Galileo (1564–1642) to which Steno refers is entitled *Discorso al Serenissimo Don Cosimo II, Gran Duca di Toscana, Intorno alle Cose che Stanno in Su L'Acqua O Che in Quella Si Muovono*. Cf. *Le Opere di Galileo Galilei, Edizione Nationale*, Firenze, Vol. IV, 1894, pp. 63–141.

a way indeed that the striæ in opposite planes ran along in the same direction, while planes adjacent to each other showed a different direction of the striæ. From the direction of the striæ it follows that the surrounding fluid was directed about every cube by a threefold movement.[1] Of these movements

P. 51. one was perpendicular to the horizon; the remaining two, parallel to the horizon, were perpendicular in relation to each other. And it is not difficult to account for this threefold movement; for while the fluid is trying to withdraw from the earth's centre, that direct movement is checked by the base of the cube, with the result that the fluid is deflected toward the narrower sides, inasmuch as the force of the ascending fluid is stronger along the wider sides and so allows no approach in that quarter; and in this way two pairs of planes are marked out by the traces of the striæ. The third pair of planes receives its striæ from that part of the fluid which passes between the cube and the fluid rebounding from the base of the cube.

4. In perfection of form; for in crystals scarcely a single one is found in whose form something is not lacking. Cubes of marcasites, however, rarely have a missing part; and the explanation is not difficult. For, since all the solid angles in the crystal, except the terminal angles, are obtuse, and the crystalline matter is added little by little to their separate planes, any given plane remains imperfect, if the adjacent planes change their shape, in just the degree that more substance is added to that one alone. Since in cubes of marcasites, however, all the solid angles are right angles, even if new matter be added to one plane only, that same plane always retains the same dimension, provided the adjacent planes do not change their form.

Various other things may be noted in the cubes of marcasites, such as cubes enclosed in cubes; the transparent matter[2]
P. 52. enveloped in the substance of the marcasite which encloses another marcasite; and other matters of this kind, which I keep for the Dissertation itself.

There are also angular bodies which are broken up into

[1] Steno was probably the first to observe that the cube surfaces of pyrite are commonly striated parallel to three intersecting edges.
[2] Not clear, because pyrite is opaque.

layers, just as rhomboidal selenites are rhomboid bodies which are broken up into other rhomboidal bodies.[1] And there are various other bodies which, although differing from the crystal in many respects, still all agree in this, that they were formed in a fluid and from a fluid. This is true also of talc, the most famous among chemical substances; so that they are by no means mistaken who believe that the solid body of talc can be resolved into a fluid body, seeing it is beyond cavil that talc was formed from a fluid. But there is no doubt that they are as far as possible astray from the truth who strive to wrench this token from it by means of fire's violence; for talc, accustomed to kindlier treatment at Nature's hands, scorning so great barbarity in lovers of beauty, by way of revenge yields to the fire-god that function of self-destruction which it keeps closed within itself.[2]

If a careful investigation of angular bodies should be begun, touching not only their composition but also their decomposition, we should soon gain a sure knowledge concerning the diversity of the motion by which the particles of both the attenuated fluid and the surrounding fluid are driven on; and this branch of physics is as essential to all for a true understanding of the workings of Nature as few they be that pursue it.

SHELLS OF MOLLUSKS

Among solids naturally enclosed in a solid none occurs more commonly, or occasions greater doubt, than the shells of mollusks. Concerning these, therefore, I shall speak at somewhat

---

[1] The reference appears to be to the cleavage of selenite.

[2] Talc was "famous" in alchemy. Compare White, *The Hermetic and Alchemical Writings of Paracelsus*, Vol. II, London, 1894, p. 383 (*A Short Lexicon of Alchemy*):

"The older alchemists have often made reference to what they term an Oil of Talc, to which they have attributed so many virtues that subsequently chemists have exerted all their power to compose it. They have calcined, purified, and sublimed the matter in question, but have met with no success. The reason is that the term was used allegorically, and that the reference was to the Oil of the Philosophers, the elixir at the white."

See also Robert Boyle, *The Usefulness of Philosophy* (edition of P. Shaw, London, 1725, Vol. I, p. 67):

"But a credible person, disciple to Cornelius Drebell, cou'd do more than this. He assur'd me, he had a way of building furnaces, wherein he, by the single force of fire, made *Venetian* talc flow; which I confess myself unable to do by the fire of a glass-house." "Talc, usually employ'd in cosmetics, is of so very difficult calcination, that eminent chymists have look'd on all calces of talcs as counterfeit." *Ibid.*, p. 158.

greater length, considering first shells taken from the sea, and then those which are dug from mountains.

Shells of every kind which at one time had a living creature enclosed in them, reveal to our perceptions the following characteristics:

1. The entire shells are themselves resolved into subdivisions, the subdivisions, again, are divided into filaments, and these filaments are reduced to two kinds differing from each other in color, composition, and place.

2. In the subdivisions of the shell the upper and lower surfaces are nothing but the ends of filaments, while the surface of the edge is the sides of those same filaments located in the edge of the subdivision.

3. The inner surface of the shell itself is identical with the inner surface of the inmost or largest subdivision, while the outer surface is composed of the outer surface of the smallest subdivision, and of the surface of all the edges of the intermediate subdivisions.

Regarding the manner in which shells on animals are formed, the following points can be clearly shown:

1. That the substance of the filaments is like the perspiration of animals, in that it is the fluid exuded through the outer surface of the animal.

2. That the form of the filaments can be produced in two ways: either in the animal's very pores, through which they are exuded, or the surface of the growing animal, having become larger than the surface of the subdivisions already hardened, separates from it, and so partly draws the viscous fluid contained between the two surfaces into filaments (a process which is common to viscous fluids), and partly adds to it by the exudation of fresh fluid, because no other substance can enter between the two surfaces.

3. That the difference of the filaments depends upon a difference of the pores by which the surface of the animal is perforated, and upon a difference of the substance which is exuded through the pores; for animals of this kind possess a twofold substance in their surface, of which the one is harder, the other

softer, and both fibrous; a careful examination of these is as illuminating as an investigation of bones.

4. That all the subdivisions, if you exclude the outermost or smallest, were produced between the outer shell and the body of the animal itself, and so have received their forms, not from themselves, but from their place; the result of this is that in the case of oysters the motion of the animal, and the amount of substance, often give rise to some diversity of form. With regard to the outermost shell there can be a doubt whether the surrounding fluid has touched the outer surface or whether it has been protected by a membrane. I should, however, believe that the last view alone is correct: (1) Because the filaments of all the rest of the subdivisions were untouched by the surrounding fluid at the time when they formed. (2) Because in prickly cockles we see that something like a membrane or skin covers the outside of the shells. But the inquiry concerns something almost outside the realm of vision, and it can be said that the filaments of the first subdivision had already hardened within the egg, since experimental knowledge proves that oysters and other testacea spring from eggs, not from decaying matter.[1]

From what has been said it is easy to explain:

1. All the diversity of hues and of spines which arouse the wonderment of many in the case of shells not only from our own land, but also from other lands; for it has no other origin than the edge of the animal enclosed in the shell. This edge, gradually growing and expanding from something exceedingly small, leaves its impress upon each margin of the subdivisions, since these margins either form from the fluid which is exuded from the outer edge of the animal, or are themselves the creature's outer edges which, like the teeth of the shark, grow

[1] Theories of spontaneous generation were common among the Greek philosophers; as, *e.g.* Anaximander, in Diels, *Fragmente der Vorsokratiker* (Zweite Auflage, Berlin, 1906), p. 17, and especially Aristotle, *de Animalium Historia*, V. 1, 3, and *de Generatione Animalium*, I. 23; III. 9, 10, and 11. Steno's friends were the first to combat them scientifically; so Harvey, *Exercitationes de Generatione Animalium* (London, 1651); Francesco Redi, *Esperienze intorno alla Generazione degl' Insetti* (Florence, 1668); Swammerdam, *Historia Insectorum Generalis* (Utrecht, 1669). See also Huxley, *Address before the British Association*, 1870, in *Lay Sermons, Addresses, and Reviews* (New York, 1877), pp. 345-378.

up anew, perhaps, in the place of the earlier edge and, like those same teeth, are gradually thrust outward.

2. The formation of pearls, not only of those which, clinging to the shells, have a form not quite round, but also of those which, after the pores in the surface of the animal have closed, acquire a round form within the pores themselves. For between the integuments of pearls and the subdivisions of shell of pearl-bearing mollusks there is merely this difference, that the filaments of the shells are located in the same plane, as it were, while the integuments of pearls have their filaments distributed over a spherical surface.

P. 56.

A happy instance of this was furnished by a pearl which, with others, I broke at your command. This pearl, although white without, enclosed within it a black body resembling a grain of pepper in both color and size; in that black body the position of the filaments tending toward the centre was very clear, and the arrangements or spheres of the same filaments could be distinguished. At the same time I saw:

1. That the excrescences on various pearls are nothing else than very small pearls enclosed by the same common crusts.

2. That many pearls of yellowish hue are imbued with a yellow color not only in the outermost surface of the sphere, but in all the inner spheres; so that it is thus no longer possible to doubt that the yellow color must be attributed to the changing fluids of the animal, and that he who seeks to wash it clear, washes an Ethiopian;[1] unless the color has either been acquired, as, for instance, the tint gained from being worn at the throat, or else was yellow in the outermost sphere only, as might be the case if, for instance, the fluids of the animal had changed when the inner spheres were being formed.

In view of these facts the mistake is apparent of those who, without a knowledge of Nature, cleverly attempt the imitation of pearls; for hardly any one could assay that feat successfully

[1] The allusion is to a fable of Æsop, who flourished about the middle of the sixth century, B.C. (Halm, *Fabulae Aesopicae*, Leipzig, 1875, XIII): 'A man bought an Æthiopian believing his color to be what it was through the neglect of his former owner. Upon taking him home, the new owner applied to him soaps of all kinds and tried to whiten him with baths of every description. But he was unable to change the color and was ready to fall sick from his toil. Characteristics remain what they were.'

unless another Lucullus should fill his aquaria[1] with pearl-bearing mollusks, and either seek in the animals themselves the methods of increasing them, or learn thence the difficulty of imitating the works of Nature. I would not deny that one can

P. 57. form, artificially, globular masses consisting of various integuments, but to arrange these integuments from a succession of filaments and unite them according to a system, upon which the natural lustre of pearls depends, this I should consider indeed most difficult.

The shells which lie buried in the earth are reduced to three classes.

The first class consists of those which are as like the shells just described as an egg is to an egg; since both the shells themselves are resolved into subdivisions, and the subdivisions into filaments; and there is the same difference and position of filaments. An examination of the shell itself proves that these shells were parts of animals at one time living in a fluid, even if marine testacea had never been seen, as will appear from the example of bivalve mussels.

At the time when bivalve mussels were formed, the substance contained within the mussel,

1. Had a smooth surface pierced with countless pores, and a twofold variation of pores.
2. Had a substance pliable and less hard than the shell itself.
3. Was in communication on the one side with the surrounding matter, on the other had no communication with it.
4. Gradually withdrew, from the side where communication with the outer matter was denied it, toward the side where it had free communication with that same matter.

---

[1] The fish-ponds of Lucullus were famous in antiquity, but our sources do not indicate that he stocked them with pearl-bearing oysters. Nor does Steno imply that he did so. The *locus classicus* is Pliny's *Natural History*, IX, 170 (54): 'Lucullus cut away a mountain near Naples at greater expense than it had cost him to build his villa. He let in the seawater, and for this reason Pompeius Magnus used to call him the Roman Xerxes. After the death of Lucullus the fish were sold for four thousand sestertia' (more than $150,000). With this may be compared Pliny, *N. H.*, VIII, 211 (52) and Plutarch, *Lucullus*, 39.

P. 58.
    5. Was able to open itself at times in proportion to the size of the angle which the hinges of the shells allow.
    6. Grew from a small to a large size.
    7. Transmitted through its own substance the matter of which the subdivisions of shell were made.

As regards the outer matter surrounding mussels: (1) If it was not wholly a fluid, at least its power of resistance was less than the power of expansion inherent in the substance within the shells. (2) It contained a fluid substance suited to the formation of the filaments of the subdivisions of shell.

All these conditions of both the inner and the outer place, which are proved by arguments and drawings in the Dissertation itself, fully show that there was an animal within the shells, and a fluid without the shells.

The second class consists of those shells which are in other respects like those just described, but differ from them in color and weight. While some are lighter, others are heavier, because the heavier shells have their pores filled with an extraneous fluid, while the pores of the former have been enlarged by the ejection of the lighter parts; I add nothing further in regard to them because they are nothing but the shells of animals, either petrified or calcined.

The third class consists of shells which in form only are similar to those just described, in other respects differing from them completely; since neither subdivisions nor filaments, much less differences of filaments, are found in them. Some of these are filled with air; others, either black or yellow in
P. 59. color, with stone; others with marble; others with crystal; and still others with other matter. The formation of all of these I account for in the following way.

When the penetrating force of fluids has dissolved the substance of a shell, the fluids have either been drained away by the earth, leaving empty spaces in the shells (which I call shells filled with air),[1] or have been changed by the addition of new matter and have filled the spaces in the shells with crystals or marble or stone, according to the diversity of the matter. And

[1] The Latin phrase is *testas aereas*; porous shells are meant.

from this source that most beautiful kind of marble which is called Nephiri[1] has its origin, which is nothing else than a deposit of the sea filled with shells of every description, in which a stony substance takes the place of the decomposed substance of the shells.

The limitation of my plan does not allow me to give an exposition of all the things which I have remarked worthy of notice in the different kinds of shells dug from the earth; wherefore, passing by other matters, I shall mention here only the following:

1. A pearl-bearing mussel, found in Tuscany, in which the pearl was clinging to the shell itself.

2. A part of an unusually large pinna in which, after the decomposition of the byssus, the color of the byssus remained in the earthy matter which had filled the shell.

3. There are shells of oysters of marvellous size in which are found several oblong cavities eaten out by worms, quite

---

[1] The term *Nephiri* is unintelligible. Neither nepheline nor nephrite, to which the word bears closest resemblance, fulfils the requirements of a marine deposit. Maar's note (*Om Faste Legemer*, p. 105) leaves the difficulty unsolved: "Vi har intetsteds kunnet finde nogen Oplysning om *Nephiri*. Professor Heiberg, til hvem vi har henvendt os, antager det for en Trykfejl for *nephriti* og formoder, at marmor nephrites er det, der nu hedder Breccia (Marmarkonglomerat). Hertil maa dog bemaerkes, at nefritisk Marmor naeppe indeholder noget, der af Steno kunde antages for Muslingeforsteninger. Tozzetti (*Reisen* I, p. 127) omtaler nefritisk Marmor fra Bygninger i Pisa, og angiver, at det i Virkeligheden er en Slags Serpentine." See also *Opera Philosophica*, Vol. II, p. 340.

It is more probable that Steno wrote *Septarium*, which was converted to *Nephiri* by a printer's error. The change is not difficult to account for palæographically. The final *um* following a vowel was usually indicated by ~ ; *S* was taken for *N* ; *ph* was an error for *pt*, and *i* for *a*. Steno's chirography was none too clear, as may be seen from the facsimile letter inserted by Wichfeld, *Erindringer om den Danske Videnskabsmand Niels Stensen, Dansk historisk Tidsskrift*, 3 Raekke, 4, opposite p. 108. That the error was not corrected in the *legenda*, p. 79, of the original edition cannot be urged against this conjecture, inasmuch as many mistakes in the text escaped notice.

Septaria are thus described by Chamberlain and Salisbury (*A College Text-book of Geology*, New York, 1909, p. 48): "Concretions sometimes develop cracks within themselves, and these may then be filled with mineral matter differing in composition or color from that of the original concretions (Fig. 28). Concretions the cracks of which have been filled by deposition from solution, are called *septaria*. They are especially abundant in some of the Cretaceous shales and clays. In not a few cases the filling of the cracks appears to have wedged segments of the original concretion farther and farther apart, until the outer surface of the septarium is made up more largely of vein-matter than of the original concretion. The development of concretions in rock is not commonly looked upon as metamorphism, but it is really a metamorphic change in the broadest sense of that term."

like those which[1] a certain kind of mussel inhabits in the rocks of Ancona, Naples, and Sicily. These cavities in the rocks, unless they were formed by insects building a nest out of mud (a thing which I can scarcely believe, since the substance of the middle of the rock, where no cavities are found, is identical with the substance of the rock containing the cavities, which are all confined to the surfaces), must have been eaten out by worms; and this view is not only commended by the surface of the cavity, but also proved by a body composed of rather thick filaments which is found in many cavities, and which answers to the cavity itself in size and shape. Surely, the cavities were made neither by mussels nor around mussels, since testacea of this kind lack the organs for gnawing, and no cavity corresponds to the shape of the shells.[2] Nor is it surprising that rocks exposed to the sea afford a resting place, in their cavities, for mussels' eggs which have been cast up by the sea, for I have not yet seen any of those cavities lacking an evident exit. But if one say that the cavities were made by a petrifying fluid which became hard around certain bodies, some cavities would have been found enveloped by that same substance on every side, and lacking an exit.

4. A shell partly destroyed on the inside, where a marble incrustation covered by various balanoids had supplied the loss of the substance eaten away; so that it is possible to infer with certainty that the shell had been left upon land by the sea, next carried down into the sea; again covered by a new deposit, and abandoned by the sea.

5. Very small eggs and helical shells hardly visible except with the aid of the microscope.

6. Pectens, helical shells, and bivalve mollusks not covered with crystal but crystalline in all their substance.

7. Various tubes of sea worms.

[1] The Florentine edition has *quos*: read *quas*.

The cavities, containing the thick filaments referred to by Steno, were probably made by the Lithophagus (Lithodomus). This Lamellibranch, of the family Mytilidae, perforates shells of the Lamellibranchs Melina, Ostrea, and Pectin, and produces a flask-shaped excavation. See von Zittel, *Grundzüge der Paläontologie, Dritte Auflage, 1 Abteilung* (München, 1910), p. 322 and fig. 632 *c*.

[2] The borer is a mussel; cf. the preceding note.

OTHER PARTS OF ANIMALS

What has been said concerning shells must also be said concerning other parts of animals, and animals themselves buried in the earth. Here belong the teeth of sharks, the teeth of the eagle-fish,[1] the vertebræ of fishes, whole fish of every kind, the crania, horns, teeth, femurs, and other bones of land animals; since all these are either wholly like true parts of animals, or differ from them only in weight and color, or have nothing in common with them except the outer shape alone.

A great difficulty is caused by the countless number of teeth which every year are carried away from the island of Malta; for hardly a single ship touches there without bringing back with it some proofs of that marvel. But I find no other answer to this difficulty than:

1. That there are six hundred and more teeth to each shark, and all the while the sharks live new teeth seem to be growing.

2. That the sea, driven by winds, is wont to thrust the bodies in its path toward some one place and to heap them up there.

3. That sharks come in shoals and so the teeth of many sharks can be left in the same place.

4. That in lumps of earth brought here from Malta,[2] besides different teeth of different sharks, various mollusks are also found, so that even if the number of teeth favors attributing their production to the earth, yet the structure of these same teeth, the abundance in each animal, the earth resembling the bottom of the sea, and the other sea objects found in the same place, all alike support the opposite view.

P. 62.

Others find great difficulty in the size of the femurs, crania, and teeth, and other bones, which are dug from the earth. But the objection, that an extraordinary size makes it necessary to conclude the size to be beyond the powers of Nature, is not of so great moment, seeing that:

---

[1] Steno's phrase is *piscis aquilae*, 'eagle-fish.' The reference is to a family of rays known scientifically as *Myliobatidae* and popularly as "eagle-rays," "devil-fishes," and "sea-devils." The teeth are flat molars, adapted for crushing hard substances.

[2] Cf. p. 211.

1. In our own time bodies of men of exceedingly tall stature have been seen.

2. It is certain that men of unnatural size existed at one time.[1]

3. The bones of other animals are often thought to be human bones.

4. To ascribe to Nature the production of truly fibrous bones is the same as saying that Nature can produce a man's hand without the rest of the man.

There are those to whom the great length of time seems to destroy the force of the remaining arguments, since the recollection of no age affirms that floods rose to the place where many marine objects are found to-day, if you exclude the universal deluge, four thousand years, more or less, before our time. Nor does it seem in accord with reason that a part of an animal's body could withstand the ravages of so many years, since we see that the same bodies are often destroyed completely in the space of a few years. But this doubt is easily answered, since the result depends wholly upon the diversity of soil; for I have seen strata of a certain kind of clay which by the thinness of their fluid decomposed all the bodies enclosed within them. I have noticed many other sandy strata which preserved whole all that was entrusted to them. And by this test it might be possible to come to a knowledge of that fluid which disintegrates solid bodies. But that which is certain, that the formation of many mollusks which we find to-day must be referred to times coincident with the universal deluge, is sufficiently shown by the following argument.

It is certain that before the foundations of the city of Rome were laid, the city of Volterra was already powerful. But in the exceedingly large stones which are found in certain places (the remains of the oldest walls) at Volterra, shells of every kind are found,[2] and not so very long ago there was hewn from the

---

[1] The belief in the existence of giants, based upon the finding of fossil bones of beasts, was widespread. See E. B. Tylor, *Researches into the Early History of Mankind* (London, 1865), pp. 314–317; *Primitive Culture*, 4th edition (London, 1903), Vol. I, p. 387.

[2] The courses of massive masonry within the impressive Porta all' Arco are of a yellow conchiliferous sandstone, called *panchina*. See Dennis, *Cities and Cemeteries of Etruria* (London, 1878), Vol. II, p. 144.

midst of the forum a stone packed full of striated shells; hence it is certain that the shells found to-day in the stones had already been formed at the time when the walls of Volterra were being built.

And in order that no one may say that the shells only have turned into stone, or that having been enclosed within the stone they have suffered no destruction from the tooth of time, we may remark that the whole hill upon which the most ancient of Etruscan cities is built, rises from the deposits of the sea, placed one above the other, and parallel to the horizon; and in these deposits many strata, not of stone, abound in

P. 64. mollusks that are real and have suffered no change at all; so it is possible to affirm that the unchanged shells which we dig from them to-day were formed three thousand and more years ago. From the founding of Rome to our own times, we reckon two thousand four hundred and twenty years and more; who will not grant that many ages elapsed from the time the first men transferred their homes to Volterra until it grew to the flourishing size it possessed at the time of the founding of Rome? And if to these centuries we add the time which intervened between the first sedimentary deposit of the hill of Volterra, and the time when that same hill was left by the sea and strangers flocked to it, we shall easily go back to the very times of the universal deluge.

The same authority of history forbids our doubting that those exceedingly large bones which are dug from the fields of Arezzo, have withstood the ravage of nineteen hundred years; for it is certain:

1. That the skulls of the pack-animals which are found there do not belong to animals of this climate, as neither do the huge femurs, and very long scapulæ, which are found in the same place.

2. It is certain that Hannibal crossed thither before he fought with the Romans at the Trasumene Lake.

3. It is certain that there were in his army African pack-animals and turret-bearing elephants of huge size.

4. It is certain that while he was coming down from the

P. 65. mountains of Fiesole a large part of the animals kept for carrying packs perished in the marshy places from the excessive floods.[1]

5. It is certain that the place whence are dug the bones under discussion, was heaped up from various strata which are full of stones rolled down by the force of torrents from the surrounding mountains; so that the manifest agreement in all details can no longer remain hidden from one who compares the character of the place and of the bones with the historical record.

PLANTS

What has been said regarding animals and their parts holds equally true of plants and the parts of plants, whether they are dug from earthy strata or lie hidden within rocky substance; for they either completely resemble actual plants and parts of plants (this kind is found rather rarely), or they differ from actual plants only in color and in weight (this kind occurs more frequently, sometimes burnt in charcoal, sometimes impregnated in a petrifying fluid), or they correspond to actual plants in

---

[1] Cuvier (*Recherches sur les Ossemens Fossiles*, 4th edition, Vol. II, Paris, 1834, p. 17) states that the skeleton of an elephant was found at Arezzo in 1663:

"C'est dans le val de Chiana, au territoire d'Arezzo, que le grand-duc Ferdinand II, ce généreux protecteur des sciences, fit déterrer un squelette entier en 1663, dont proviennent encore, selon Targioni (Tozzeti), une partie des os conservés à Florence, et dont paraissent avoir parlé Stenon et Boccone."

Steno's explanation is a naïve attempt to account for the presence, in very great numbers, of the fossil remains of elephants belonging to the Pleistocene period. Says Cuvier (*op. cit.*, p. 16): "Quand on passe de l'Etat de l'Eglise en Toscane, en suivant le Tibre, le Clanis ou Chiana et l'Arno, les os d'éléphans deviennent de plus en plus nombreux. Le val de Chiana, le val d'Arno et les vallées particulières qui y aboutissent, en contiennent d'étonnantes quantités."

Further references are: Forsyth Major, *Considerazioni sulla Fauna dei Mammiferi pliocenici e post-pliocenici della Toscana* in *Atti di Società Toscana di Scienze Naturale in Pisa*, Vol. I (1875), pp. 7-40, 223-245; III (1877), pp. 207-227; *Mammalian Fauna of the Val d'Arno* in *The Quarterly Journal of the Geological Society of London*, Vol. XLI (1885), pp. 1-8; Depéret, *Evolution of Tertiary Mammals, and the Importance of Their Migrations* in *The American Naturalist*, Vol. XLII (1908), pp. 109-114, 166-170, 303-307.

It may be interesting to note, apropos of Steno's "historical record," that both Livy and Polybius agree that Hannibal had only one elephant by the time he reached Arezzo. According to Eutropius (III. 8) and Polybius (III. 42), Hannibal entered Italy with thirty-seven elephants. Polybius states (III. 74) that all except one perished from the extreme cold immediately after the battle of the Trebia (218 B.C.), and Livy (XXI. 56) remarks that almost all (*prope omnis*) were overcome. In the attempt to cross the Apennines, a detail of the campaign which is not mentioned by Polybius and is a source of confusion in Livy, seven of these succumbed (Livy, XXI. 58). And when Hannibal reached Arezzo in the early spring of 217 B.C. (XXII. 2), Livy represents him as riding the sole survivor.

form only; of this last kind there is a great abundance in various places.

Regarding the first two classes there cannot be the least doubt that they were at one time actual plants, since the structure of their very bodies compels this view, and the character of the place whence they are dug does not oppose it. They who hold, in opposition, that the earth which had been carried over into houses in process of time changed into wood, cannot affirm this except of the surface of the earth enclosing the wood, where the earth, having become dry in time, and turned to dust, has brought to light the wood enclosed within it. Neither do the metallic filaments found in the pores of the same wood militate against our view, since I myself have drawn from the earth a trunk attesting its plant nature by the knots of its branches and by its bark, whose fissures had been filled with mineral matter. Hence, furthermore, it might throw no little light upon the lore of minerals if an investigation were made of the wood, and the place of the wood to determine what they could have contributed to the formation of minerals. Under the name of bitumen come many things which the channels of the fibres and the ashes of the burnt portions prove to be nothing but charcoal.

P. 66.

A greater difficulty is occasioned by the third class of plants, or the forms of plants marked upon stones, since we find forms of this kind in hoar-frost, in the mercury tree,[1] in various volatile salts, in a white substance [2] soluble in water, which not only forms in glass vessels on their inner surfaces but sometimes rises from the middle of the vessel into the air. But to one who duly weighs all considerations nothing will be found to be inconsistent with the views expressed; for the forms of plants inscribed on stones are reducible to two classes. Some of these forms are imprinted only on the surface of the clefts, which I would readily acknowledge to have been produced without an actual plant, although not without a fluid; others appear not only on the surface of the clefts but spread their branches everywhither throughout the substance of the stone. Hence it follows that at the time when a plant of the second type was

---

[1] The metallic crystals produced by mercury in a solution containing silver.
[2] Ammonium chloride, sal ammoniac.

being produced, whether this took place after the manner of other plants, or in the fashion of the mercury tree, the substance of the stone had not yet laid aside the character of a fluid; this fact, again, is further assured not only by the softer consistency of the stone, but also by the angular bodies common in the dendrite of Elba, such as form only in a free fluid. But what need is there of other proofs, when experience itself speaks? I have seen various moist places, not only those exposed to the sun, but also underground, where, on account of water flowing by, a rock growing to moss and other plants was being covered with new moss of a different kind.

Hitherto I have reviewed the principal bodies whose place of finding has for many afforded no clue to the place of their production; and at the same time I have hinted how, from that which is perceived, a definite conclusion is formed in regard to that which cannot be perceived.

THE DIFFERENT CHANGES WHICH HAVE TAKEN PLACE IN TUSCANY

In what way the present condition of any thing discloses the past condition of the same thing, is above all other places clearly manifest in Tuscany; inequalities of surface observed in its appearance to-day contain within themselves plain tokens of different changes, and these I shall review in inverse order, proceeding from the last to the first.

1. At one time the inclined plane A [Pl. XI, fig. 20] was in the same plane with the higher, horizontal plane B, and the end of the same plane A thus raised, as also the end of the higher, horizontal plane C, were continuous, whether the lower, horizontal plane D was in the same plane with the higher horizontal planes B, C, or another solid body existed there, supporting the exposed sides of the higher planes. Or, what is the same thing, in the place where to-day rivers, swamps, sunken plains, steeps, and planes inclined between sand hills are seen, all was once level, and at that time all the waters, both of rains and of springs, were flooding that plain, or had opened for themselves underground channels beneath it; at any rate, there were cavities under the upper strata.

2. At the time when the plane B, A, C [Pl. XI, fig. 21]

was being formed, and other planes under it, the entire plane B, A, C, was covered with water; or, what is the same thing, the sea was at one time raised above sand hills, however high.

3. Before the plane B, A, C [Pl. XI, fig. 22] was formed, the planes F, G, I [Pl. XI, fig. 23] had the same position which they now hold; or, what is the same thing, before the strata of the sand hills were[1] formed, deep valleys existed in the same places.

4. At one time the inclined plane I [Pl. XI, fig. 23] appeared in the same plane with the horizontal planes F and G, and either the exposed sides of the planes I and G were continuous, or another solid existed there, supporting the exposed sides when the planes were being formed; or, what is the same thing, where valleys are seen to-day between the plane summits of the highest mountains, there was at one time a single continuous plane under which huge caverns had been formed before the downfall of the upper strata.

5. When the plane F, G [Pl. XI, figures 24 and 25] was being formed, a watery fluid lay upon it; or, what is the same thing, the plane summits of the highest mountains were at one time covered with water.

Six distinct aspects of Tuscany[2] we therefore recognize, two when it was fluid, two when level and dry, two when it was broken; and as I prove this fact concerning Tuscany by inference from many places examined by me, so do I affirm it with reference to the entire earth, from the descriptions of different places contributed by different writers. But in order that no one may be alarmed by the novelty of my view, in a few words I shall set forth the agreement of Nature with Scripture by reviewing the chief difficulties which can be urged regarding the different aspects of the earth.

In regard to the first aspect of the earth Scripture and Nature agree in this, that all things were covered with water[3]; how and when this aspect began, and how long it lasted, Nature says not, Scripture relates. That there was a watery

[1] The alluvial deposits of the valley.
[2] This summary takes up the figures in inverse order, figures 25, 24, etc.
[3] See *Genesis*, I. 1–7.

fluid, however, at a time when animals and plants were not yet to be found, and that the fluid covered all things, is proved by the strata of the higher mountains, free from all heterogeneous material. And the form of these strata bears witness to the presence of a fluid, while the substance bears witness to the absence of heterogeneous bodies. But the similarity of matter and form in the strata of mountains which are different and distant from each other, proves that the fluid was universal. But if one say that the solids of a different kind contained in those strata were destroyed in course of time, he will by no means be able to deny that in that case a marked difference must have been noticed between the matter of the stratum and the matter which percolated through the pores of the stratum, filling up the spaces of the bodies which had been destroyed. If, however, other strata which are filled with different bodies are, in certain places, found above the strata of the first fluid, from this fact nothing would follow excepting that above the strata of the first fluid new strata were deposited by another fluid, whose matter could likewise have refilled the wastes of the strata left by the first fluid. Thus we must always come back to the fact that at the time when those strata of matter unmixed, and evident in all mountains, were being formed, the rest of the strata did not yet exist, but that all things were covered by a fluid free from plants and animals and other solids. Now since no one can deny that these strata are of a kind which could have been produced directly by the First Cause, we recognize in them the evident agreement of Scripture with Nature.

Concerning the time and manner of the second aspect of the earth, which was a plane and dry, Nature is likewise silent, Scripture speaks. As for the rest Nature, asserting that such an aspect did at one time exist, is confirmed by Scripture, which teaches us that the waters welling from a single source overflowed the whole earth.[1]

When the third aspect of the earth, which is determined to have been rough, began, neither Scripture nor Nature makes plain. Nature proves that the unevenness was great, while Scripture makes mention of mountains[2] at the time of the

---

[1] See *Genesis*, 2. 10-14.   [2] See *Genesis*, 7. 19-20.

flood. But when those mountains, of which Scripture in this connection makes mention, were formed, whether they were identical with mountains of the present day, whether at the beginning of the deluge there was the same depth of valleys as there is to-day, or whether new breaks in the strata opened new chasms to lower the surface of the rising waters, neither Scripture nor Nature declares.

The fourth aspect, when all things were sea, seems to cause more difficulty, although in truth nothing difficult is here presented. The formation of hills from the deposit of the sea bears witness to the fact that the sea was higher than it is now, that too not only in Tuscany but in very many places distant enough from the sea, from which the waters flow toward the Mediterranean; nay, even in those places from which the waters flow down into the ocean. Nature does not oppose Scripture in determining how great that height of the sea was, seeing that:

1. Definite traces of the sea remain in places raised several hundreds of feet above the level of the sea.

2. It cannot be denied that as all the solids of the earth were once, in the beginning of things, covered by a watery fluid, so they could have been covered by a watery fluid a second time, since the changing of the things of Nature is indeed constant, but in Nature there is no reduction of anything to nothing. But who has searched into the formation of the innermost parts of the earth, so that he dare deny that huge caverns may exist there, filled sometimes with a watery fluid, sometimes with a fluid akin to air?

3. It is wholly uncertain what the depth of valleys at the beginning of the deluge was; reason, however, may urge that in the first ages of the world smaller cavities had been eaten out by water and fire, and that in consequence not so deep breaks of strata followed from this cause; while the highest mountains of which Scripture speaks were the highest of those mountains which were in existence at that time, not of those which we see to-day.

4. If the movement of a living being can bring it to pass that places which have been overwhelmed with waters are arbitrarily made dry, and are again overwhelmed with waters,

why should we not voluntarily grant the same freedom and the same powers to the First Cause of all things?

In regard to the time of the universal deluge, secular history is not at variance with sacred history, which relates all things in detail. The ancient cities of Tuscany, of which some were built on hills formed by the sea, put back their birthdays beyond three thousand years; in Lydia, moreover, we come nearer to four thousand years: so that it is possible thence to infer that the time at which the earth was left by the sea agrees with the time of which Scripture speaks.[1]

As regards the manner of the rising waters, we could bring forward various agreements with the laws of Nature. But if some one say that in the earth the centre of gravity does not always coincide with the centre of the figure, but recedes now on one side, and now on the other, in proportion as subterranean cavities have formed in different places, it is possible to assign a simple reason why the fluid, which in the beginning covered all things, left certain places dry, and returned again to occupy them.

The universal deluge may be explained with the same ease if a sphere of water, or at least huge reservoirs, be conceived around a fire in the middle of the earth; thence, without the movement of the centre, the pouring forth of the pent-up water could be derived. But the following method also seems to me to be very simple, whereby both a lesser depth of the valleys and a sufficient amount of water are obtained without taking into account the centre, or figure, or gravity. For if we shall have conceded (1) That by the slipping of fragments of certain strata, the passages were stopped through which the sea penetrating into hollow places of the earth sends forth the water to bubbling springs; (2) That the water undoubtedly enclosed in the bowels of the earth, was, by the force of the known subterranean fire in part driven toward springs, and in part forced up into the air through the pores of the ground which had not

P. 73.

---

[1] Steno accepted the chronology of Archbishop Usher, which assigned the creation of the world to the year 4004 B.C. In this connection, see also J. Woodward, *An Essay Toward a Natural History of the Earth and Terrestrial Bodies, Especially Minerals*, etc., London, first edition, 1695; J. Arbuthnot, *An Examination of Dr. Woodward's Account of the Deluge. . . . With a Comparison between Steno's Philosophy and the Doctor's in the Case of Marine Bodies Dug out of the Earth*, London, 1697.

yet been covered with water; that, moreover, the water which not only is always present in the air but also was mixed with it in the manner previously described, fell in the form of rain; (3) That the bottom of the sea was raised through the enlarging of subterranean caverns; (4) That the cavities remaining on the surface of the earth were filled with earthy matter washed from the higher places by the constant falling of rains; (5) That the very surface of the earth was less uneven, because nearer to its beginnings — if we shall have granted all this, we shall have admitted nothing opposed to Scripture, or reason, or daily experience.

What happened on the surface of the earth while it was covered with water, neither Scripture nor Nature makes clear; this only can we assert from Nature, that deep valleys were formed at that time. This is (1) because the cavities, made larger by the force of subterranean fires, furnished room for greater downfalls; (2) because a return passage had to be opened for the waters into the deeper parts of the earth; P. 74. (3) because to-day, in places far from the sea are seen deep valleys filled with many marine deposits.

As for the fifth aspect, which revealed huge plains after the earth had again become dry, Nature proves that those plains existed, and Scripture does not gainsay it. For the rest, whether the entire sea presently receded, or whether, indeed, in the course of ages new chasms opening afforded opportunity for disclosing new regions, it is possible to determine nothing with certainty, since Scripture is silent, and the history of nations regarding the first ages after the deluge is doubtful in the view of the nations themselves, and thought to be full of myths. This, indeed, is certain, that a great amount of earth was carried down every year into the sea (as is easily clear to one who considers the size of rivers, and their long courses through inland regions, and the countless number of mountain streams, in short, all the sloping places of the earth), and that the earth thus carried down by rivers, and added day by day to the shore, left new lands suited for new habitations.

This is in fact confirmed by the belief of the ancients, in accordance with which they called whole regions the gifts of rivers[1] of like name, as also by the traditions of the Greeks,[2] since they relate that men, descending little by little from the mountains, inhabited places bordering on the sea that were sterile by reason of excessive moisture, but in course of time became fertile.

The sixth aspect of the earth is evident to the senses; herein the plains left by the waters, especially by reason of erosion, and at times through the burning of fires, passed over into various channels, valleys, and steep places. And it is not to be wondered at that in the historians there is no account as to when any given change took place. For the history of the first ages after the deluge is confused and doubtful in secular

---

[1] In Homer the river Nile (*Odyssey* IV. 477) is ὁ Αἴγυπτος and the country (*Odyssey* XVII. 448) is ἡ Αἴγυπτος. Herodotus (II. 5) calls Aegypt 'the gift' of the Nile, and Plato (*Timaeus*, 22 D) represents an Egyptian priest as saying to Solon: 'And from this calamity (*i.e.* periodic destruction) the Nile, which is our never-failing savior, saves and frees us.' Cf. Strabo, *Geography*, C. 36 (I. 2, 29).

[2] The tradition is explained at length in dialogue in Plato's *Laws*, 677–682 B:

'*Athenian.* Do the ancient traditions seem to you to contain any truth?

Kleinias. What traditions?

Ath. The traditions that many destructions of mankind were occasioned by deluges and diseases and many other things, as a result of which only a small remnant of the human race was left.

Kl. Every one believes all that.

Ath. Come then. Let us think that one of many such destructions was once occasioned by a deluge.

Kl. What are we to think about it?

Ath. That those who then escaped the destruction would only be some mountain shepherds, small sparks of the human race preserved on the mountain tops.

Kl. Clearly. . . .

Ath. After this they came together in greater numbers and increased the size of their cities, and turned to husbandry, first at the foot of the mountains. . . .

Ath. Homer also disclosed the form following the second, saying that the third arose thus. For he says, He founded Dardania, since holy Ilios had not yet been built on the plain, the city of mortal men, but they still dwelt at the foot of Ida with its many springs (*Iliad*, XX. 216–218). . . .

Ath. Now Ilios, we say, was built when men had come down from the heights into a large, fair plain, on a low hill watered by many rivers which descended from Ida.'

We may compare also Plato's *Timaeus*, 22 C; *Statesman*, 270; *Critias*, 109 D.

A summary of the passage in the *Laws* is given by Strabo in his *Geography*, C, 592, 593 (XIII. 1, 25):

'Plato conjectures that three forms of political commonwealth were established after the deluges. The first was upon the mountain tops, a simple and savage affair, composed of folk who feared the waters that still flooded the plains. The second was on the foothills, composed of folk who regained their courage little by little, since the plains were beginning to dry. The third was in the plains. One might mention a fourth and a fifth, and possibly more, but the last was on the sea-coast and in the islands, after all fear of a deluge had vanished.'

writers; as the ages passed, moreover, they felt constrained to celebrate the deeds of distinguished men, not the wonders of Nature. Nevertheless the records, which ancient writers mention, of those who wrote the history of the changes which occurred in various places, we do not possess. But since the authors whose writings have been preserved report as marvels almost every year, earthquakes, fires bursting forth from the earth, overflowings of rivers and seas, it is easily apparent that in four thousand years[1] many and various changes have taken place.

Far astray, therefore, do they wander, who criticize the many errors in the writings of the ancients, because they find there various things inconsistent with the geography of to-day. I should be unwilling to put credence in the mythical accounts of the ancients; but there are in them also many things to which I would not gainsay belief. For in those accounts I find many things of which the falsity rather than the truth seems doubtful to me. Such are the separation of the Mediterranean Sea[2] from the western ocean; the passage from the Mediterranean into the Red Sea; and the submersion of the island Atlantis.[3] The description of various places in the journeys of Bacchus, Triptolemus, Ulysses, Æneas, and of others, may be true, although

P. 76. it does not correspond with present day facts. Of the many changes which have taken place over the whole extent of Tuscany embraced between the Arno and the Tiber, I shall adduce evident proofs in the Dissertation itself; and although the time, in which the individual changes occurred, cannot be determined, I shall nevertheless adduce those arguments from the history of Italy, in order that no doubt may be left in the mind of anyone.

And this is the succinct, not to say disordered, account of the

---

[1] In order to account for the evolution of earth features within the time limit imposed by his belief in Usher's chronology of creation, Steno is compelled to adopt a theory of violent catastrophes in nature.

[2] See p. 210, note 1.

[3] Plato, *Timaeus*, 25, C, D: 'But later, when violent earthquakes and deluges occurred, in a single day and night of misfortune, all your military power in a body sank into the earth, and in a like manner the island Atlantis sank and disappeared in the sea. For this reason the sea in that region is even now impassable and impenetrable because a shoal of mud (reading κάρτα βραχέος instead of βαθέος) forms a barrier which was caused by the sinking island.' Compare also *Critias*, 108 E, ff.

principal things which I had decided to set forth in the Dissertation, not only with greater clearness but also with greater fulness, adding a description of the places where I have observed each thing.

P. 77. Let Vincentius Viviani examine this work, and report whether there is anything in it which is contrary to the Catholic Faith or to good morals.

  Vinc. de Bardis, *Vicar General of Florence.*

Most Illustrious and Most Reverend Sir:

Having seen the new and admirable *Prodromus* of the most distinguished Steno, the *Dissertation Concerning a Solid Naturally Contained Within a Solid*, or rather, if you prefer, of the whole of Physics, and having recognized in it a perfectly sincere manifestation of the Catholic faith and of good morals, as in the very candid author, I have indeed thought the same worthy of being entrusted to type on this, the thirtieth day of August, 1668.

  Vincentius Viviani.

Let it be printed servatis servandis.

  Vinc. de Bardis, *Vicar General of Florence.*

The seventh day of December, 1668.

Let Franciscus Redi, Consultor of the Holy Office at Florence, examine it and report.

Fr. Jacobus Tosini, *Vicar General of the Holy Office at Florence.*

Most Reverend Father:

The *Prodromus* of the very learned and expert Nicolaus Steno's *Dissertation Concerning a Solid Naturally Contained* P. 78. *Within a Solid*, adorned in the highest degree with sound and noble learning, I have seen, and I have judged it worthy of printing.

  Franciscus Redi.

With the foregoing attestation let it be printed at Florence this day, the thirteenth of December, 1668.

Fr. Joseph Tamagninus, *Chancellor of the Holy Office at Florence.*

Gio. Federighi, *Senator and Auditor of the Holy Apostolic Chamber, and through him, Benedetto Gori.*

EXPLANATION OF THE FIGURES

INASMUCH as the brevity of my hurried writing has left not a few things insufficiently explained, especially where the treatment concerns angular bodies and the strata of the earth, in order to afford some sort of remedy for that defect, I have decided to add here the following figures, chosen from very many others.

[PLATE IX]

The first thirteen figures, intended to illustrate the angular bodies of crystal, fall into two classes.

The first class contains seven varieties of a plane in which the axis of the crystal lies.

In figures 1, 2, and 3, the axes of the parts, of which the body of the crystal is composed, form a straight line; but there is an intermediate prism, which is lacking in Figure 1, appears rather short in 2, longer in 3.

In Figure 4, the axes of the parts which make up the body of the crystal do not form a straight line.

Figures 5 and 6 belong to the class of those which I could present in countless numbers to prove that in the plane of the axis both the number and the length of the sides are changed in various ways without changing the angles; that various cavities are left in the very middle of the crystal, and that various layers are formed. Figure 7 in the plane of its axis shows how both the number and the length of the sides are sometimes increased in various ways, sometimes diminished, from the new crystalline matter placed above the planes of the pyramids.

The second class contains six varieties of base of planes.[1]

In Figures 8, 9, 10, and 11, there are only six sides; with this difference, nevertheless, that in Figure 8 all the sides are equal, while

[1] The cross section parallel to the basal pinacoid.

PLATE IX.

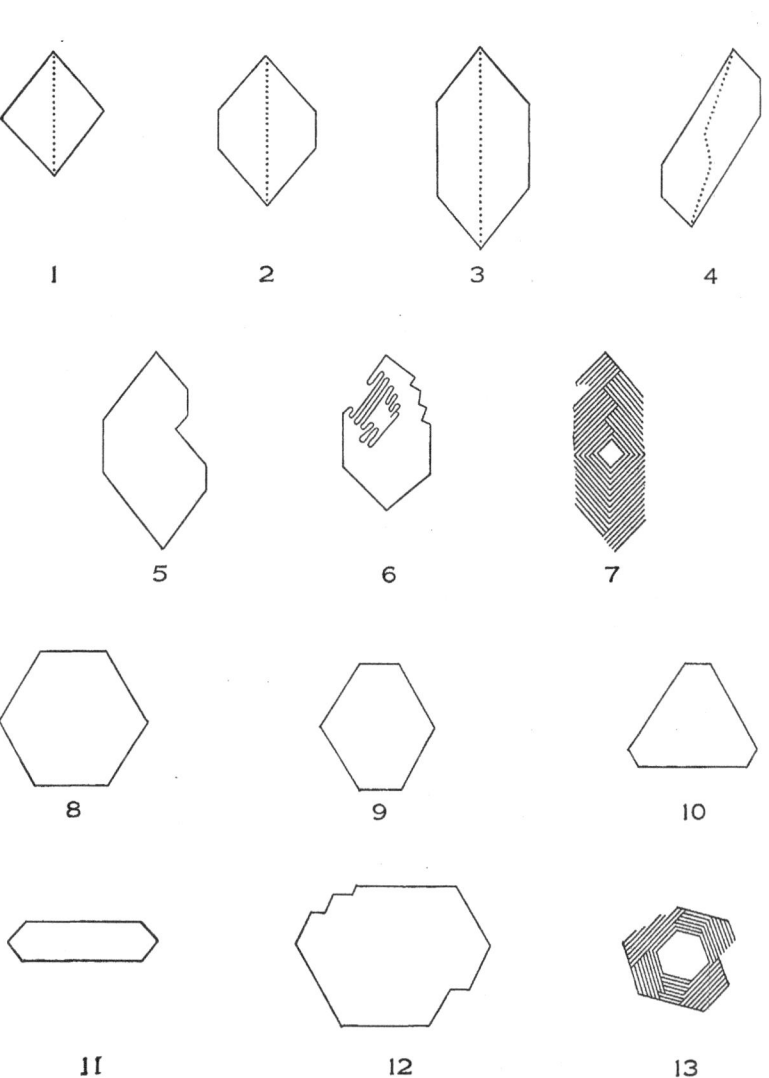

STENO'S FIGURES 1–13, IN EXACT SIZE.

in Figures 9 and 11 not all, but only the opposite sides, are equal; in Figure 10, any given opposite sides are unequal.

In Figure 12 the plane of the base, which ought to be a hexagon, is bounded by twelve sides.

Figure 13 shows how sometimes the length and number of the sides are changed in various ways without changing the angles, on the plane of the base, while new crystalline matter is being placed upon the planes of the pyramids.

[PLATE X]

The six following figures explain two different kinds of angular bodies of iron.

Figures 14, 15, 16 serve to illustrate those angular bodies of iron which are enclosed by twelve planes; Figure 14, in fact, shows all the twelve planes spread out into a single plane, six of these being triangular and brilliant, the remaining six pentagonal and striated. Figure 15 is the plane of the base of the same body. Figure 16 is the plane of the axis of the same body.

Figures 17, 18, and 19 serve to illustrate those angular bodies of iron which are bounded by thirty planes.

Figure 17 shows all the thirty planes spread out into a single plane; of these six planes are pentagonal and brilliant, twelve triangular and also brilliant, six triangular and striated, six oblong quadrilaterals and brilliant.

Figure 18 is the plane of the base of the same body.

Figure 19 is the plane of the axis of the same body.

PLATE X.

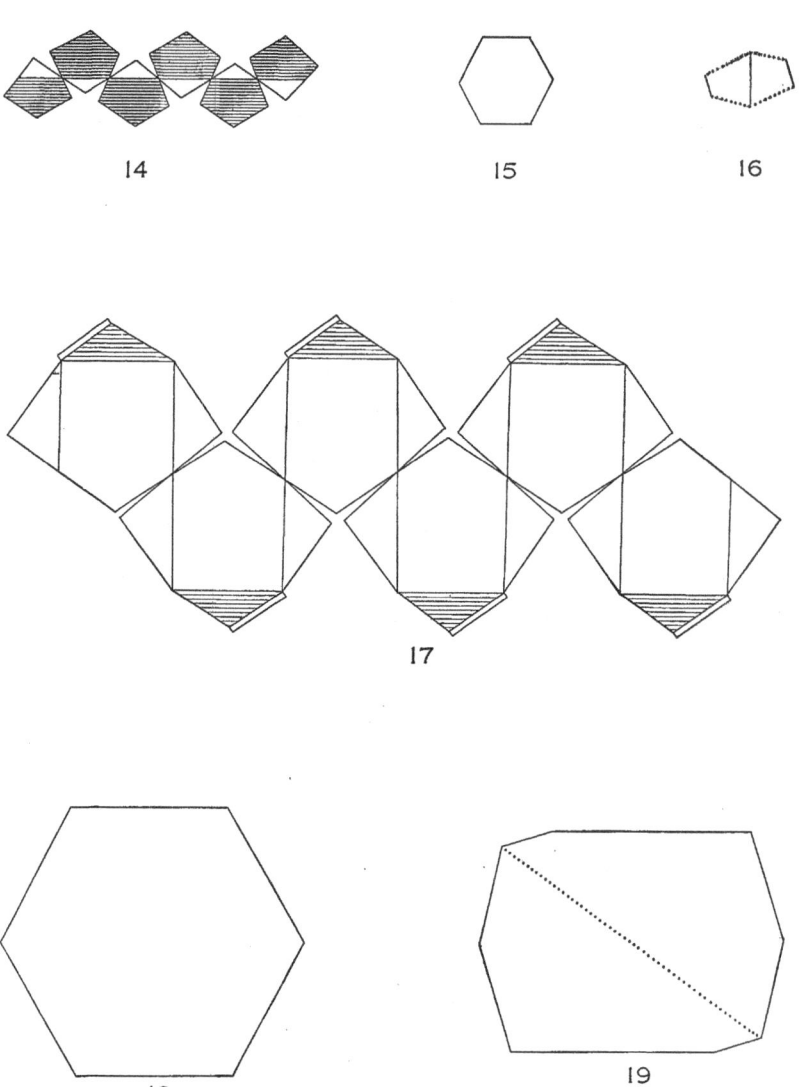

STENO'S FIGURES 14–19, IN EXACT SIZE EXCEPT 17.

[PLATE XI]

The last six figures, while they show in what way we infer the six distinct aspects of Tuscany from its present appearance, at the same time serve for the readier comprehension of what we have said about the earth's strata. The dotted lines represent the sandy strata of the earth, so called from the predominant element, although various strata of clay and rock are mixed with them; the rest of the lines represent strata of rock, likewise named from the predominant element, although other strata of a softer substance are sometimes found among them. In the Dissertation itself I have explained the letters of the figures in the order in which the figures follow one another: here I shall briefly review the order of change.

Figure 25 shows the vertical section of Tuscany at the time when the rocky strata were still whole and parallel to the horizon.

Figure 24 shows the huge cavities eaten out by the force of fires or waters while the upper strata remained unbroken.

Figure 23 shows the mountains and valleys caused by the breaking of the upper strata.

Figure 22 shows new strata, made by the sea, in the valleys.

Figure 21 shows a portion of the lower strata in the new beds destroyed, while the upper strata remain unbroken.

Figure 20 shows the hills and valleys produced there by the breaking of the upper sandy strata.

PLATE XI.

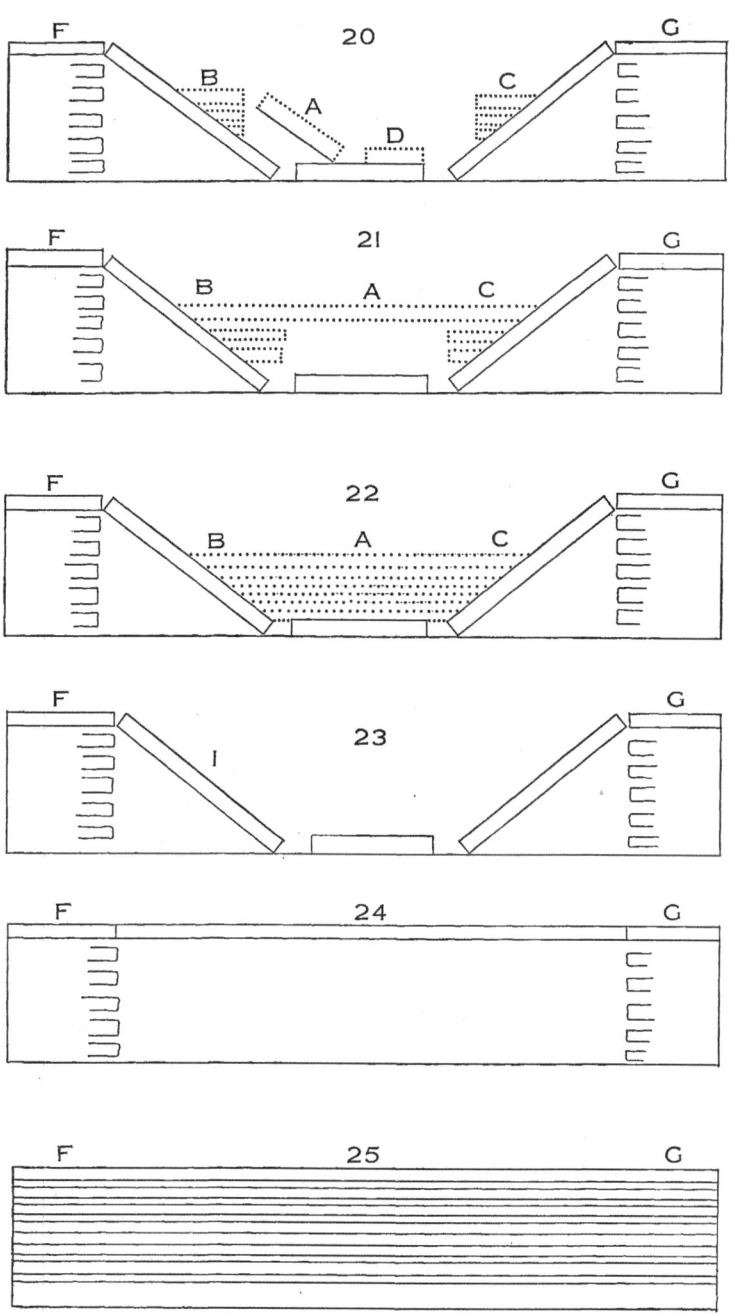

STENO'S FIGURES 20–25, IN EXACT SIZE.

INDEX

Abel, 211, n. 1; 224, n. 1.
Academics, 213, n. 2; 214.
Accademia del Cimento, founding of, 180; Steno member of, 209.
Acosta, de, 236, n. 2.
Accretions, 225; inorganic, 232, n. 2; to marcasites, 247.
Æneas, 269.
Æschylus, 205, n. 3.
Æsop, 252, n. 1.
Aëtites, 224, n. 1.
Africa, land bridge from, 174.
Agate, 224, 225.
Agent, as form or idea, 216; determining motion, 216; universal, 217.
Agricola, George, on glossopetræ, 211, n. 1; on mountain formation, 232, n. 2; on gnomes, 232, n. 2; on divining rod, 236, n. 1.
Air, in breathing, 221; Hippocrates' theory of, 223; explosion of, 231; in crystals, 238.
Albertus Magnus, on glossopetræ, 211, n. 1.
Alchemy, 249, n. 2.
Alimentary canal, 221.
All Souls' Day, 180, n. 3.
Alps, 234, n. 2.
Alum, 243; feathery, 225.
Amethysts, 225.
Amianthus, 224.
Ammon, marine deposits at temple of, 210, n. 1.
Ammonium chloride, 261, n. 2.
Amniotic fluid, 220.
Amsterdam, Steno's arrival in, 176, 183, 208, n. 1.
Anaximander, 251, n. 1.
Ancients, 267, 269; on fossils, 210 f.
Ancona, 256.
Angelis, de, on Steno's degree, 176, n. 1.
Angles, law of constancy of interfacial, 171; of crystals, 237, 248; of hematite, 245; of marcasites, 248.
Angular bodies, 220; with meaning of crystals, 225, 226, 244; of iron, 244 f.; place and production of, 244; planes of hematite, 245; striation of, 244 f.
Animals, 216; fluids in, 221; formation of, 221; parts of, 257; place of growth in, 220; shells of, 218.
Antimony, 218.
Apennines, 260, n. 1.

Aqua fortis, 243, n. 1; regia, 243, n. 1.
Aquaria, 253.
Arabian Gulf, 210, n. 1.
Arbuthnot, 266, n. 1.
Arezzo, fossils at, 259.
Aristotle, 251, n. 1.
Armenia, salt lakes in, 210, n. 1.
Arno, extinct animals in valley of, 174, 260, n. 1; 269.
Arnolfini, Lavinia Felice Cenanni, 180.
Artaxerxes, 210, n. 1.
Arteries, 222.
Ashes, in strata, 229, 232.
Atlantic, level of, 210, n. 1.
Atlantis, 269.
Atoms, 216.
Axes, of crystals, 237, 272 f.

Bacchus, 269.
Bacon, Francis, 181, n. 3; 196, 205, n. 3.
Balanoids, 256.
Banks, Sir Joseph, 197.
Bardis, de, 271.
Barlæus, Caspar, 232, n. 1.
Bartholin, Thomas, 175, 176 f., 207, n. 3; 211, n. 1.
Baudry, Paul, 205, n. 3.
Beaumont, Elie de, 201.
Bezoar, 224.
Bitumen, 229, 232, 261.
Bladder, 221.
Blaes, Gerard, 176, 177, 178.
Blondel, 181.
Blood, circulation of, 222.
Boccone, 260, n. 1.
Bodies, produced naturally, 215; not produced by earth and rocks, 218.
Bohemia, 232, n. 2.
Bologna, 186.
Bones, broken, 225; fibrous, 258; fossil taken for human, 258; enclosed in solids, 218.
Boni, 232, n. 2.
Borch, Ole, see Borrichius.
Borrichius, 175, 178.
Bossuet, 180.
Boyle, Robert, 198, 199, 199, n. 1; 200, n. 1; 232, n. 1; 236, n. 1; 241, n. 2; 243, n. 1; 249, n. 2.
Brazil, 232.

Breccia, 255, n. 1.
Bruno, Giordano, 169.
Burial, Chinese, 236.
Byssus, 255.
Byzantium, 210, n. 1.

Calculi, 221.
Callus, 225.
Canals, 222.
Candolle, 181, n. 3.
Capellini, Giovanni, 186.
Capillaries, 221, 222, 223.
Casserius, 176.
Cause, First, 264, 266.
Caverns, 234, 265, 267.
Caves, 234.
Cavities, in body, 222; in crystals, 238; in earth, 230; in rocks, 238, 256; in shells, 255.
Chalcedony, 224.
Chamberlain, 255, n. 1.
Charcoal, 229, 260.
Chemists, 216.
Chéreau, 175, n. 2; 176, n. 1.
Chiana, 260, n. 1.
Chinese, 236.
Chorion, 220.
Christian V, 183.
Cicero, 176, n. 4; 205, n. 3.
Cinnabar, 218.
Clanis, 260, n. 1.
Clays, 255, n. 1.
Cleavage, of selenites, 249, n. 1.
Cockles, 210, n. 1; 251.
Concretions, 226, n. 1; 255, n. 1.
Consani, Vincenzo, 185.
Copenhagen, 175, 176, 185.
Copper, 218, 225, 246.
Cosimo III, 182, 183, 185.
Cracks, in concretions, 255, n. 1; in strata, 231, 232.
Crania, 257.
Creation, strata at, 228; chronology of, 269.
Crete, 210, n. 1.
Cribration, 222, 223, 224.
Crusts, 228, 229.
Crystals, angles of, 237; cause of, 242; columnar form of, 171; faces of, 171; figures of, 272 ff.; formation of, 237 ff.; growth of, 171, 238 f.; hardening of, 218, 237; hues of, 240, 243; enclosed in solid, 218; law of interfacial angles of, 171; lustre of, 240; of mountains, 225; movement in, 242; of niter, 171, 219; nucleus of, 171, 172; orientation of molecules in, 171; parts of, 237; phantom, 171; planes of, 237, 239; prisms of, 237; pyramids of, 237; means quartz, 218, n. 1; 220, n. 1; 237, n. 1; in shells, 254; surface of, 239.
Crystalline matter, 238, 239, 240, 241.
Crystallography, 171.
Cubes, of marcasites, 247; truncated, 245.
Cuvier, 260, n. 1.
Cybotus, 232, n. 2.
Cynics, 213, n. 2; 214.

Damigeron, 224, n. 1.
Dardania, 268, n. 2.
Dati, Carlo, 180.
Davis, Robert, 198.
Deluge, Universal, 169, 174, 258, 264, 265, 266, 268.
Democritus, 205.
Demons, 232, n. 2.
Demosthenes, 224, n. 1.
Dendrites, 225, 262.
Denmark, 232, n. 2.
Dennis, 258, n. 2.
Depéret, 260, n. 1.
Deposits, alluvial, 263; marine, 172, 226.
Descartes, 170, 179, 228.
Devil-fishes, 257, n. 1.
Diamonds, 225, 246.
Diels, 251, n. 1.
Dissertation, 181, 208, 215, 248, 254, 269, 276.
Divination, 236.
Divining rod, 173, 236, n. 1.
Dog's head, dissection of, 177.
Dragon, 236.
Drebell, Cornelius, 249, n. 2.
Ducts, lymphatic, 222.

Eagle, 224, n. 1.
Eagle-fish, teeth of, 257.
Eagle-rays, 257, n. 1.
Eaglestone, 224, 224, n. 1.
Earth, productivity of, 212, 216, 217.
Earthquakes, 173, 232, n. 2; 235, 269.
Egypt, 210, n. 1; 268, n. 1.
Elba, 236, 244, n. 1; 262.
Elements, chemical, 216; four, 216.
Elephants, 174, 259.
Elixir, 249, n. 2.
Epicurus, 213, n. 2.
Erasistratus, 222, n. 2.
Eratosthenes, 210, n. 1.
Erosion, 268.
Ethiopian, hue of, 252.
Eutropius, 260, n. 1.
Euxine, outlet of, 210, n. 1.
Exhalations, 235.
Eyelids, 221.

INDEX

Eyes, 221, 242.
Eysson, 177.

Fabronius, 232, n. 2.
Faces, of crystal, 239.
Fat, 225.
Femurs, 257, 259.
Federighi, 271.
Ferdinand II, collection of minerals, 246, n. 1; death of, 182; dedication to, 205; fossils found by, 260, n. 1; patron of Steno, 169, 179, 180, 207, 211, n. 1.
Fibres, of muscle, 222, 224, 225; of plants, 221, 225.
Fiesole, 259.
Figures, explanation of, 272 ff.
Filaments, of mineral, 224; in rocks, 256; in receiver, 242; of shells, 174, 250 f., 254.
Filings, of iron, 241.
Fire, agency in breaking up crystals, 243; agency in mountain formation, 232, n. 2; proof of in strata, 229; subterranean, 232, 266, 267.
Fish, 253, 257.
Fissures, containing crystals, 238; containing minerals, 218.
Floods, 211, 229, 258, 260, 269.
Florence, Geologists' Congress in honor of Steno, 186; fossils from Arezzo in, 260, n. 1; Steno's arrival in, 183; topography of, 173.
Flourens, 182, n. 3.
Fluid, aqueous, 219; common, 222, 223; defined, 222; differing from solid, 171; external, 220, 221: internal, 220, 221, 223; peculiar, 223; penetrating, 217; permeating, 214, 221; petrifying, 256.
Fontenelle, 208, n. 2.
Foramina, of body, 221.
Fossils, elephants, 260, n. 1; origin, 170, 173, 200; from sea, 208, 208, n. 2; 210, 211.
Frederik III, 181, 182, 208, n. 1.
Furnaces, 249, n. 2.
Fürstenberg, von, 184.

Galen, 222, n. 2.
Galileo, 247, n. 2.
Gases, subterranean, 231.
Generation, spontaneous, 251, n. 1.
Genesis, 263, n. 3; 264, n. 1, 2.
Germany, silver in, 246.
Gerra, 210, n. 1.
Giants, 258, n. 1.
Gibraltar, land passage at, 210, n. 1.
Glands, 207, 222.
Glass, distinguished from crystal, 171; formation of, 243.

Glossopetræ Melitenses, 211, 211, n. 1.
Gnomes, 232, n. 2.
God, 216.
Gold, 243, n. 1.
Golias, Jacob, 178.
Goniometer, 171.
Gori, 271.
Granites, 225.
Grafting, 215.
Grass, in strata, 228.
Greeks, 268.
Griffenfeldt, Count, 178, 183.
Growth, of crystals, 238; of mountains, 232.
Gypsum, 218, n. 1.

Hamburg, 184.
Hannibal, pack animals of, 174, 259, 260, n. 1.
Hannover, 182, n. 1.
Hardening, of crystals, 237; of solids, 218.
Harvey, 222, n. 4; 251, n. 1.
Heart, 207.
Heat, 232.
Heiberg, 255, n. 1.
Hellespont, 210, n. 1.
Hematite, 244, n. 1.
Heraclitus, 205, n. 3.
Herodotus, 268, n. 1.
Hesperian, sinking of, 198, n. 3.
Hills, formation of, 232, n. 2; 263.
Hippocrates, 223.
Hoar-frost, 261.
Holland, religious tolerance in, 178, 180.
Homer, 268, n. 1.
Hooke, Robert, 173, 197, 201.
Hoover, 232, n. 2; 236, n. 1.
Horizontality, of strata, 172, 230.
Horne, Van, 178.
Horns, 257.
Hues, of crystals, 243; of pearls, 252; of shells, 251, 254.
Hughes, 176, n. 1.
Humboldt, A. von, 182, n. 3.
Hungary, 232, n. 2.
Huxley, 182, n. 3; 251, n. 1.

Ice, making, 215.
Ida, 268, n. 2.
Idols, 236.
Iliad, 268, n. 2.
Ilios, 268, n. 2.
Incrustations, growth of, 220, 224; on solids, 226.
India, 246.
Innsbruck, 182.
Insects, 256.
Interstices, in body, 214.

Intestines, 221.
Iron, cubes of, 225; of Elba, 236; growth of, 236; n. 3; figures of crystals of, 273; filings of, 241, 242.
Irradiations, 237.
Italy, 269.

Jacobæus, Matthias, 178.
Johann Friedrich, Duke of Hannover, 183, 184.

Kidneys, 222, n. 2.
Kircher, Athanasius, 180, 182, n. 2; 234, n. 1; 236, n. 1.
Knorr, 211, n. 1.

Laboulbène, 224, n. 2.
Labyrinth, 206.
Lake Sirbonis, 210, n. 1.
Lamellibranch, 256, n. 1.
Lapilli, 231.
Lapis lazuli, 218.
Laws, of nature, 215.
Layers, deposition of, 226, 227; mineral layers in rivers, 228; of rock, 226.
Lefebure, 205, n. 3.
Leibnitz, 182, n. 1; 211, n. 1.
Lernæan Hydra, 206.
Leyden, 178.
Libya, connected with Europe, 210, n. 1.
Limonite, 236, n. 3.
Linnæus, 187.
Lithodomus, 256, n. 1.
Lithophagus, 256, n. 1.
Littré, 243, n. 1.
Liver, 222, n. 2.
Livy, 260, n. 1.
Load-stone, *see* Magnet.
Lucca, 180.
Lucullus, 253.
Lungs, 222, n. 2.
Lustre, in crystals, 240.
Lydia, 266.
Lymphatic ducts, 222.

Maar, 175, n. 1; 178, n. 1, 2; 179, n. 2; 182, n. 2; 188 ff., 195, 197, 198, n. 2, 3; 202, 206, n. 1; 207, n. 5; 211, n. 1; 222, n. 1; 222, n. 3; 223, n. 1; 224, n. 2; 226, n. 2; 232, n. 1, 2; 234, n. 1; 236, n. 1, 2; 246, n. 3; 255, n. 1.
Magalotti, Lorenzo, 180.
Magnet, 170; filings about, 241, 242; lines of force, 171.
Maillet, de, 169.
Major, Forsythe, 260, n. 1.

Malta, Bartholin's journey in, 211, n. 1; stones from, 211, n. 1; teeth from, 257.
Manni, 184.
Marbles, 225.
Marcasites, cubes of, 225; formation of crystals of, 247 ff.; hardening in, 218; enclosed in solid, 218; = pyrites, 218, n. 1.
Maria Flavia del Nero, 180.
Marine deposits, 172, 210.
Matiana, 210, n. 1.
Matrix, of crystals, 238.
Matter, constitution of, 216; surrounding mussels, 254.
Maurits, Count Jan, 232, n. 1.
Maximilian Heinrich, 184.
Maxims of Morality, 214.
Medici, Leopold de', 180, 209, n. 1.
Mediterranean, 210, n. 1; 265, 269.
Medulla, 225.
Melina, 256, n. 1.
Membrane, dividing, 222.
Menstruum, 243, 243, n. 1.
Mercury, 218, 243, n. 1; tree, 261, 262.
Metals, growth of, 232, n. 2.
Metamorphism, 255, n. 1.
Microscope, 256.
Mineral, replacements of, 173; repositories of, 235 ff.; veins of, 173.
Minerals, formation of, 261; enclosed in bodies, 218; origin of, 235 f.
Miners, 236.
Mines, 232, n. 2.
Molars, 257, n. 1.
Molecules, orientation of, 171.
Mollusks, bivalve, 256, 257, 258; petrified, 225, 226; produced by Nature, 217; shells of, 249 ff.; structure of, 173.
Monardes, 246, n. 3.
Montanari, 232, n. 2.
Moss, 262.
Motion, artificial, 215; of crystalline matter, 241; determination of, 214 f., 216; first cause of, 215; of particles in fluids, 249; principle of, perception of, 214.
Mountains, causes of, 173; chains of, 234; crests of, 234; crystals of, 219; under deluge, 264; growth of, 232, 232, n. 2; origin of, 231 ff.; overthrown, 234; overwhelmed, 263; of scripture, 265; types of, 173; upheaval of, 228.
Mount Casius, 210, n. 1.
Mouth, 221.
Movements, in crystal, 242; of fluid forming marcasites, 248.
Munster, 184.
Murano, 182.

INDEX

Murchison, 234, n. 2.
Muscle, fibres, 222.
Muscles, Steno's work on, 208 ; substance of, 222.
Musgrave, George, 232, n. 1.
Mussels, 241 ; pearl-bearing, 255 ; substance in bivalve, 253.
Myliobatidæ, 257, n. 1.
Myths, 267, 269.
Mytilidæ, 256, n. 1.

Naples, 182, 253, n. 1; 256.
Natural Questions, solution of, 210, 213.
Nature, Laws of, 214 ; products of, 215, 216, 217 ; unknown, 217.
Needham, 176, n. 2.
Nepheline, 255, n. 1.
Nephiri, 255, 255, n. 1.
Nephrite, 255, n. 1.
Nile, 268, n. 1.
Nilsdatter, Anna, 178, n. 3 ; 207, n. 4.
Niter, crystals of, 219.
Nose, 221.

Odyssey, 180, 268, n. 1.
Œsophagus, 221.
Oldenburg, Henry, 197, 199 ff.
Onyx, 224.
Orange, 220.
Organs, excretory, 221.
Ostrea, 256, n. 1.
Ovum, 222.
Oysters, 210, n. 1; 241, 251, 255.

Palissy, Bernard, 208, n. 2; 224, n. 1.
Panchina, 258, n. 2.
Parenchymata, 222, 224.
Parotid duct, 176.
Particles, added to a solid, 226 f.; changeable, 216; growth of, 220, 221; motion of, 214, 216.
Passageways, for things issuing from the earth, 234 f.
Paulli, Jacob Henry. 177.
Paulli, Simon, 176.
Pearls, formation of, 252; imitation of, 252; in mussels, 255; structure of, 174.
Pectens, 256.
Pectin, 256. n. 1.
Pedersen, 175, 178, n. 3.
Peduncle, 220.
Pelusium, 210, n. 1.
Penetrating fluid, 241.
Pepys, Samuel, 197, n. 1.
Peripatetics, 213, n. 2; 214.
Peru, 236.
Phrygia, 210, n. 1.

Pine cones, in strata, 228.
Pinna, 255.
Pisa, 255, n. 1.
Piso, Willem, 232, n. 1.
Pitti Palace, 182, 246, n. 2.
Place, definition of term, 219; place of crystal formation, 237; of production, 217.
Plains, 267.
Planes, of crystals of hematite, 244, 245; of crystals of quartz, 237, 239; of diamonds, 246; of marcasites, 248.
Plants, anatomy of, 221; fibrous parts of, 225; fossil, 226, 260 f.; formation of, 221; enclosed in solid, 218; metallic, 225; place of, 219; roots of, 220.
Plato, 222, n. 4; 268, n. 2; 269, n. 3.
Pleistocene, 260, n. 1.
Plenkers, 175, 175, n. 1; 178, n. 1; 178, n. 3; 179, n. 2; 180 f.; 180, n. 2; 183, 184, 185, 188, 211, n. 1.
Pliny, 211, n. 1; 224, n. 1; 253, n. 1.
Plutarch, 253, n. 1.
Poisoner, 224, n. 1.
Polybius, 260, n. 1.
Pompeius, 253, n. 1.
Pontus, 210, n. 1.
Pope Innocent XI, 184.
Pores, 221, 223, 253.
Porta all' Arco, 258, n. 2.
Prague, 182.
Pregnancy, 224, n. 1.
Prism, 237.
Prodromus, bibliography of, 194 ff.; date of composition, 181; division of subjects, 170; geometrical form of, 170; science of, 169; scope of, 181 f.; use of word, 181, n. 3; translation of, 205 ff.
Production, of diamonds, 246; of hematite, 244; of substances, 216.
Propontis, 210, n. 1.
Pumice stone, in strata, 229.
Pyramids, of crystal, 237, 239; of hematite, 245.
Pyrites, 218, n. 1; 248, n. 1.

Quartz, 218, n. 1; chemical constitution of, 219, n. 1; = crystals, 220, n. 1; 237, n. 1; 245, n. 3.
Quicksand, 235.

Rabelais, 205, n. 3.
Rains, 229, 232, 267.
Ray, 232, n. 1; 257, n. 1.
Receiver, 242.
Redi, Francesco, 180, 181, 251, n. 1; 271.
Red Sea, passage to, 269.

Refraction, of light, 171; in glass, 243.
Replacements, 220, 225.
Repositories, of minerals, 235 f.
Reservoirs, 234, 266.
Retort, 242.
Rivers, agency in mountain formation, 232, n. 2; changed by earthquake, 235; deposits of, 267; gifts of, 268.
Rocks, 172.
Rome, 182, 229, 258, 259.
Rome de l'Isle, 171.
Roots, of plants, 220.
Rose, Valentine, 224, n. 1.
Rosnel, de, 236, n. 2 ; 246, n. 3.
Rush, in strata, 228.

Salisbury, 255, n. 1.
Salivary duct, 177.
Salt, deposit of, 228.
Salt-water lakes, 210, n. 1.
Sand, layers of, 228.
Sandstone, 258, n. 2.
San Lorenzo, 185, 186.
Santa Maria Nuova, 179, 180.
Sap, of earth, 232, n. 2.
Sardinia, 210, n. 1.
Savignani, Emilio, 180.
Scallop shells, 210, n. 1.
Scandinavia, 184.
Scapulæ, 259.
Schmidt, Waldmar, 186 f.
Schumacher, Peter, *see* Griffenfeldt.
Schwerin, 185.
Scripture, 263, 264, 265, 266, 267.
Sea, 173, 228, 265.
Sea-devils, 257, n. 1.
Secretions, 223, 226, n. 1.
Sediments, 172, 206, 220, 227, 229.
Selenites, 218, 218, n. 1 ; 249.
Seneca, 206, n. 2 ; 213, n. 2.
Septarium, 255, n. 1.
Serpentine, 255, n. 1.
Shark, teeth of, 206, 207, 211, 211, n. 1; 251, 257.
Sheep's head, dissection of, 177.
Shells, crystalline, 218 ; in earth, 253 ; helical, 256 ; of marine animals, 226 ; of mollusks, 173, 250 f. ; petrified, 218 ; porous, 254.
Ships, in deposits, 228.
Sicily, 210, n. 1 ; 256.
Silver, 218, 246.
Skeletons, 260, n. 1.
Skin, 221.
Slipping, of strata, 231.
Solid, addition to, 220; contained within solid, 170, 208 ; differs from fluid, 214 ; dug from earth, 226 ; hardening of, 218 ; production of, 209, 218, 220, 224, 226.
Solon, 268, n. 1.
Soul, world, 216; agency of, 217.
Speculation, 170.
Spinoza, Baruch, 178, 184, n. 1.
Spleen, 222, n. 1.
Springs, 229, 235, 266.
Steno, life of, 175 ff.; writings of, 188 ff.; bibliography on, 202 f.
Stichman, Johannes, 207, n. 4.
Stiermark, 232, n. 2.
Stoics, 213, n. 2 ; 214.
Stokes, 225, n. 1.
Stomach, 221.
Stones, 235 f.
Storms, 229.
Strabo, 210, n. 1 ; 268, n. 1 ; 268, n. 2.
Strata, alterations of, 234; deposition of, 172, 227 ff. ; fissures in, 173 ; horizontality of, 172; manner and place of production, 219 ; order of, 173; slipping of, 266 ; thrusts of, 173 ; in Tuscany, 262 ff., 273 ; variation in character of, 172.
Strato, 210, n. 1.
Striation, in crystals, 239, 240 ; in hematite, 244, 246 ; in marcasites, 247, 248.
Substance, crystalline, 225 ; place and manner of production, 170 ; tenuous, 216.
Sulphur, 232.
Sun, 216, 217.
Swammerdam, Jan, 178, 251, n. 1.
Sylvius, 177, 178, 179.

Talc, 249, 249, n. 2.
Tamagninus, 271.
Targioni, 260, n. 1.
Taygetus, 232, n. 2.
Teeth, of sharks, 211, 257; eagle-fish, 257; from Malta, 257.
Telliamed, *see* de Maillet, 169.
Tendons, 225.
Testacea, 251, 253, 256.
Theatrum Anatomicum, 183.
Thera, 232, n. 2.
Therasia, 232, n. 2.
Thévenot, 178, 179.
Thrusts, 231.
Tiber, 260, n. 1, 269.
Time, evolution of, 258.
Titian, 205, n. 3.
Titopolis, 184.
Tolerance, religious tolerance in Holland, 178.
Tongue stones, 211, n. 1.

Tools, of miners, 236.
Tosini, 271.
Tozzetti, 255, n. 1; 260, n, 1.
Trachea, 221.
Traditions, of early civilization, 268, n. 2.
Trasumene Lake, 259.
Trebia, 260, n. 1.
Trees, in strata, 228.
Trigautius, 236, n. 1.
Triptolemus, 269.
Truth, proverb of Truth in a well, 205, n. 3.
Tuff, 220.
Tuscany, geological changes in, 170, 209, 234, n. 2; 262 ff., 269, 276; fossils in, 260, n. 1; 265, 266; pearl-bearing mussels from, 255.
Tylor, 258, n. 1.

Ulysses, 269.
Umbilical vessels, 220.
Urethra, 221.
Usher, chronology of, 174, 266, n. 1; 269, n. 1.
Uterus, 221, 222.

Valleys, 232, n. 2; 265, 267.
Valmont-Bomare, 224, n. 2.
Vapors, subterranean, 229.
Vegetatio, 232, n. 2.
Veins, 222; in rocks, 225.
Vertebræ, of fishes, 257.
Vienna, 182.
Vinci, Leonardo da, 169, 173.
Virtus formativa, 211, n. 1.

Viscera, 225.
Vitriol, 243.
Viviani, Vincenzo, 169, 179, 180, 181.
Volcanoes, 172.
Volterra, shells in, 174; walls of, 258, 259.

Walchs, 211, n. 1.
Water, agency in mountain formation, 232, n. 2; in crystals, 238; deposits of turbid water, 219; issuing from earth, 234.
Weld, 197, n. 1.
Well, proverb of Truth in, 205, n. 3.
Wells, 235.
Wharton, 176, n. 5; 177.
White, 249, n. 2.
Wichfeld, 175, n. 1; 178, n. 1; 179, n. 2; 255, n. 1.
Willis, 179.
Wind, agency in mountain formation, 232, n. 2; breaking from mountain, 234 f.
Winslow, Jacques Bénigne, 178, n. 4.
Winter, 197.
Woodward, 266, n. 1.
Woodworth, 198.
Worms, 221, 255, 256.
Wren, 197, n. 1.

Xanthus, 210, n. 1.
Xerxes, 253, n. 1.

Zittel, von, 182, 204, 256, n. 1.

University of Michigan Studies

HUMANISTIC SERIES

General Editors: FRANCIS W. KELSEY and HENRY A. SANDERS

Size, 22.7 × 15.2 cm. 8°. Bound in cloth

VOL. I. ROMAN HISTORICAL SOURCES AND INSTITUTIONS. Edited by Professor Henry A. Sanders, University of Michigan. Pp. viii + 402. $2.50 net.

CONTENTS

1. THE MYTH ABOUT TARPEIA: Professor Henry A. Sanders.
2. THE MOVEMENTS OF THE CHORUS CHANTING THE CARMEN SAECULARE: Professor Walter Dennison, Swarthmore College.
3. STUDIES IN THE LIVES OF ROMAN EMPRESSES, JULIA MAMAEA: Professor Mary Gilmore Williams, Mt. Holyoke College.
4. THE ATTITUDE OF DIO CASSIUS TOWARD EPIGRAPHIC SOURCES: Professor Duane Reed Stuart, Princeton University.
5. THE LOST EPITOME OF LIVY: Professor Henry A. Sanders.
6. THE PRINCIPALES OF THE EARLY EMPIRE: Professor Joseph H. Drake, University of Michigan.
7. CENTURIONS AS SUBSTITUTE COMMANDERS OF AUXILIARY CORPS: Professor George H. Allen, University of Cincinnati.

VOL. II. WORD FORMATION IN PROVENÇAL. By Professor Edward L. Adams, University of Michigan. Pp. xvii + 607. $4.00 net.

Vol. III. LATIN PHILOLOGY. Edited by Professor Clarence Linton Meader, University of Michigan. Pp. vii + 290. $2.00 net.

Parts Sold Separately in Paper Covers:

Part I. THE USE OF IDEM, IPSE, AND WORDS OF RELATED MEANING. By Clarence L. Meader. Pp. 1–111. $0.75.

Part II. A STUDY IN LATIN ABSTRACT SUBSTANTIVES. By Professor Manson A. Stewart, Yankton College. Pp. 113–78. $0.40.

Part III. THE USE OF THE ADJECTIVE AS A SUBSTANTIVE IN THE DE RERUM NATURA OF LUCRETIUS. By Dr. Frederick T. Swan. Pp. 179–214. $0.40.

Part IV. AUTOBIOGRAPHIC ELEMENTS IN LATIN INSCRIPTIONS. By Professor Henry H. Armstrong, Drury College. Pp. 215–86. $0.40.

THE MACMILLAN COMPANY

Publishers　　　64-66 Fifth Avenue　　　New York

University of Michigan Studies — *Continued*

Vol. IV. Roman History and Mythology. Edited by Professor Henry A. Sanders. Pp. viii + 427. $2.50 net.

Parts Sold Separately in Paper Covers:

Part I. Studies in the Life of Heliogabalus. By Dr. Orma Fitch Butler, University of Michigan. Pp. 1–169. $1.25 net.

Part II. The Myth of Hercules at Rome. By Professor John G. Winter, University of Michigan. Pp. 171–273. $0.50 net.

Part III. Roman Law Studies in Livy. By Professor Alvin E. Evans, Washington State College. Pp. 275–354. $0.40 net.

Part IV. Reminiscences of Ennius in Silius Italicus. By Dr. Loura B. Woodruff. Pp. 355–424. $0.40 net.

Vol. V. Sources of the Synoptic Gospels. By Rev. Dr. Carl S. Patton, First Congregational Church, Columbus, Ohio. Pp. xiii + 263. $1.30 net.

Size, 28 × 18.5 cm. 4to.

Vol. VI. Athenian Lekythoi with Outline Drawing in Glaze Varnish on a White Ground. By Arthur Fairbanks, Director of the Museum of Fine Arts, Boston. With 15 plates, and 57 illustrations in the text. Pp. viii + 371. Bound in cloth. $4.00 net.

Vol. VII. Athenian Lekythoi with Outline Drawing in Matt Color on a White Ground, and an Appendix: Additional Lekythoi with Outline Drawing in Glaze Varnish on a White Ground. By Arthur Fairbanks. With 41 plates. Pp. x + 275. Bound in cloth. $3.50 net.

Vol. VIII. The Old Testament Manuscripts in the Freer Collection. By Professor Henry A. Sanders, University of Michigan. With 9 plates showing pages of the Manuscripts in facsimile. Pp. viii + 357. Bound in cloth. $3.50 net.

Parts Sold Separately in Paper Covers:

Part I. The Washington Manuscript of Deuteronomy and Joshua. With 3 folding plates. Pp. vi + 104. $1.25.

Part II. The Washington Manuscript of the Psalms. With 1 single plate and 5 folding plates. Pp. viii + 105–357. $2.00 net.

THE MACMILLAN COMPANY

Publishers   64–66 Fifth Avenue   New York

University of Michigan Studies—*Continued*

Vol. IX. THE NEW TESTAMENT MANUSCRIPTS IN THE FREER COLLECTION. By Professor Henry A. Sanders, University of Michigan.

    Part I. THE WASHINGTON MANUSCRIPT OF THE FOUR GOSPELS. With 5 plates. Pp. vii + 247. Paper covers. $2.00 net.

    Part II. THE WASHINGTON FRAGMENTS OF THE EPISTLES OF PAUL. (*In Preparation.*)

Vol. X. THE COPTIC MANUSCRIPTS IN THE FREER COLLECTION. By Professor William H. Worrell, Hartford Seminary Foundation.

    Part I. A FRAGMENT OF A PSALTER IN THE SAHIDIC DIALECT. The Coptic Text, with an Introduction, and with 6 plates showing pages of the Manuscript and Fragments in facsimile. Pp. xxvi + 112. $2.00 net.

Vol. XI. CONTRIBUTIONS TO THE HISTORY OF SCIENCE. (*Parts I and II ready.*)

    Part I. ROBERT OF CHESTER'S LATIN TRANSLATION OF THE ALGEBRA OF AL-KHOWARIZMI. With an Introduction, Critical Notes, and an English Version. By Professor Louis C. Karpinski, University of Michigan. With 4 plates showing pages of manuscripts in facsimile, and 25 diagrams in the text. Pp. vii + 164. Paper covers. $2.00 net.

    Part II. THE PRODROMUS OF NICOLAUS STENO'S LATIN DISSERTATION ON A SOLID BODY ENCLOSED BY PROCESS OF NATURE WITHIN A SOLID. Translated into English by Professor John G. Winter, University of Michigan, with a Foreword by Professor William H. Hobbs. With 7 plates. Pp. 165–283. Paper covers. $1.30 net.

    Part III. VESUVIUS IN ANTIQUITY. Passages of Ancient Authors, with a Translation and Elucidations. By Francis W. Kelsey. Illustrated.

Vol. XII. STUDIES IN EAST CHRISTIAN AND ROMAN ART.

    Part I. EAST CHRISTIAN PAINTINGS IN THE FREER COLLECTION. By Professor Charles R. Morey, Princeton University. With 13 plates (10 colored) and 34 illustrations in the text. Pp. xii + 87. Bound in cloth. $2.50 net.

    Part II. A GOLD TREASURE OF THE LATE ROMAN PERIOD FROM EGYPT. By Professor Walter Dennison, Swarthmore College. (*In Press.*)

THE MACMILLAN COMPANY
Publishers   64–66 Fifth Avenue   New York

University of Michigan Studies — *Continued*

Vol. XIII. Documents from the Cairo Genizah in the Freer Collection. Text, with Translation and an Introduction by Professor Richard Gottheil, Columbia University. (*In Preparation.*)

SCIENTIFIC SERIES

Size, 28 × 18.5 cm. 4°. Bound in cloth

Vol. I. The Circulation and Sleep. By Professor John F. Shepard, University of Michigan. Pp. x + 83, with an Atlas of 83 plates, bound separately. Text and Atlas, $2.50 net.

Vol. II. Studies on Divergent Series and Summability. By Professor Walter B. Ford, University of Michigan. Pp. xi + 193. $2.50.

University of Michigan Publications

HUMANISTIC PAPERS

Size, 22.7 × 15.2 cm. 8°. Bound in cloth

Latin and Greek in American Education, with Symposia on the Value of Humanistic Studies. Edited by Francis W. Kelsey. Pp. x + 396. $1.50.

CONTENTS

The Present Position of Latin and Greek, the Value of Latin and Greek as Educational Instruments, the Nature of Culture Studies.

Symposia on the Value of Humanistic, particularly Classical, Studies as a Preparation for the Study of Medicine, Engineering, Law and Theology.

A Symposium on the Value of Humanistic, particularly Classical, Studies as a Training for Men of Affairs.

A Symposium on the Classics and the New Education.

A Symposium on the Doctrine of Formal Discipline in the Light of Contemporary Psychology.

The Menaechmi of Plautus. The Latin Text, with a Translation by Joseph H. Drake, University of Michigan. Pp. xi + 130. Paper covers. $0.60 net.

THE MACMILLAN COMPANY

Publishers   64–66 Fifth Avenue   New York

Handbooks of Archaeology and Antiquities

Edited by PERCY GARDNER and FRANCIS W. KELSEY

THE PRINCIPLES OF GREEK ART

By PERCY GARDNER, Litt.D., Lincoln and Merton Professor of Classical Archaeology in the University of Oxford.

Makes clear the artistic and psychological principles underlying Greek art, especially sculpture, which is treated as a characteristic manifestation of the Greek spirit, a development parallel to that of Greek literature and religion. While there are many handbooks of Greek archaeology, this volume holds a unique place.

New Edition. Illustrated. Cloth, $2.50

GREEK ARCHITECTURE

By ALLAN MARQUAND, Ph.D., L.H.D., Professor of Art and Archaeology in Princeton University.

Professor Marquand, in this interesting and scholarly volume, passes from the materials of construction to the architectural forms and decorations of the buildings of Greece, and lastly, to its monuments. Nearly four hundred illustrations assist the reader in a clear understanding of the subject.

Illustrated. Cloth, $2.25

GREEK SCULPTURE

By ERNEST A. GARDNER, M.A., Professor of Archaeology in University College, London.

A comprehensive outline of our present knowledge of Greek sculpture, distinguishing the different schools and periods, and showing the development of each. This volume, fully illustrated, fills an important gap and is widely used as a text-book.

Illustrated. Cloth, $2.50

GREEK CONSTITUTIONAL HISTORY

By A. H. J. GREENIDGE, M.A., Late Lecturer in Hertford College and Brasenose College, Oxford.

Most authors in writing of Greek History emphasize the structure of the constitutions; Mr. Greenidge lays particular stress upon the workings of these constitutions. With this purpose ever in view, he treats of the development of Greek public law, distinguishing the different types of states as they appear.

Cloth, $1.50

GREEK AND ROMAN COINS

By G. F. HILL, M. A., of the Department of Coins and Medals in the British Museum.

All the information needed by the beginner in numismatics, or for ordinary reference, is here presented. The condensation necessary to bring the material within the size of the present volume has in no way interfered with its clearness or readableness.

Illustrated. Cloth, $2.25

GREEK ATHLETIC SPORTS AND FESTIVALS

By E. NORMAN GARDINER, M.A., Sometime Classical Exhibitioner of Christ Church College, Oxford.

With more than two hundred illustrations from contemporary art, and bright descriptive text, this work proves of equal interest to the general reader and to the student of the past. Many of the problems with which it deals — the place of physical training, games, athletics, in daily and national life — are found to be as real at the present time as they were in the far-off days of Greece.

Illustrated. Cloth, $2.50

ON SALE WHEREVER BOOKS ARE SOLD

THE MACMILLAN COMPANY

Publishers   64-66 Fifth Avenue   New York

Handbooks of Archaeology and Antiquities — *Continued*

ATHENS AND ITS MONUMENTS

By CHARLES HEALD WELLER, of the University of Iowa.

This book embodies the results of many years of study and of direct observation during different periods of residence in Athens. It presents in concise and readable form a description of the ancient city in the light of the most recent investigations. Profusely illustrated with Half-tones and Line Engravings.

Illustrated. Cloth, $4.00

THE DESTRUCTION OF ANCIENT ROME

By RODOLFO LANCIANI, D.C.L., Oxford; LL.D., Harvard; Professor of Ancient Topography in the University of Rome.

Rome, the fate of her buildings and masterpieces of art, is the subject of this profusely illustrated volume. Professor Lanciani gives us vivid pictures of the Eternal City at the close of the different periods of history.

Illustrated. Cloth, $1.50

ROMAN FESTIVALS

By W. WARDE FOWLER, M.A., Fellow and Sub-Rector of Lincoln College, Oxford.

This book covers in a concise form almost all phases of the public worship of the Roman state, as well as certain ceremonies which, strictly speaking, lay outside that public worship. It will be found very useful to students of Roman literature and history as well as to students of anthropology and the history of religion.

Cloth, $1.50

ROMAN PUBLIC LIFE

By A. H. J. GREENIDGE, Late Lecturer in Hertford College and Brasenose College, Oxford.

The growth of the Roman constitution and its working during the developed Republic and the Principate is the subject which Mr. Greenidge here set for himself. All important aspects of public life, municipal and provincial, are treated so as to reveal the political genius of the Romans in connection with the chief problems of administration.

Cloth, $2.50

MONUMENTS OF THE EARLY CHURCH

By WALTER LOWRIE, M.A., Late Fellow of the American School of Classical Studies in Rome, Rector of St. Paul's Church, Rome.

Nearly two hundred photographs and drawings of the most representative monumental remains of Christian antiquity, accompanied by detailed expositions, make this volume replete with interest for the general reader and at the same time useful as a hand-book for the student of Christian archaeology in all its branches.

Illustrated. Cloth, $1.50

MONUMENTS OF CHRISTIAN ROME

By ARTHUR L. FROTHINGHAM, Ph.D., Sometime Associate Director of the American School of Classical Studies in Rome, and formerly Professor of Archaeology and Ancient History in Princeton University.

"The plan of the volume is simple and admirable. The first part comprises a historical sketch; the second, a classification of the monuments." — *The Outlook.*

Illustrated. Cloth, $2.25

ON SALE WHEREVER BOOKS ARE SOLD

THE MACMILLAN COMPANY

Publishers   64-66 Fifth Avenue   New York